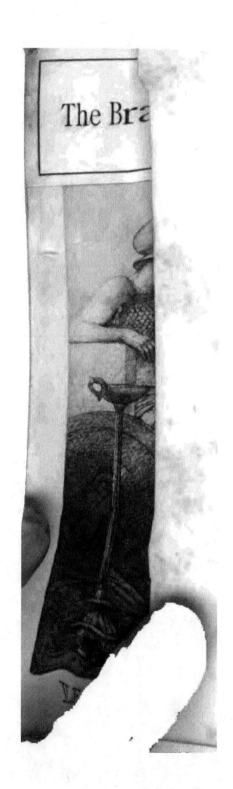

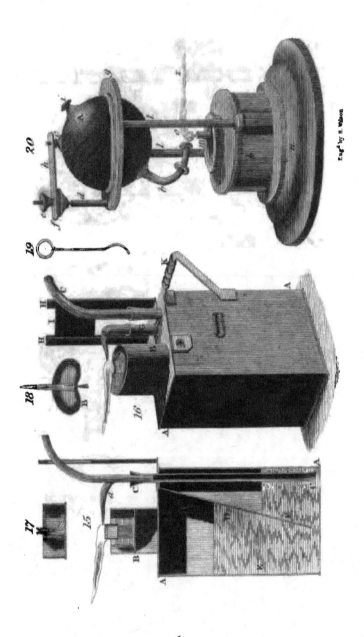

A
PRACTICAL TREATISE
ON
THE USE OF
THE
BLOWPIPE.

BY
John Griffin.

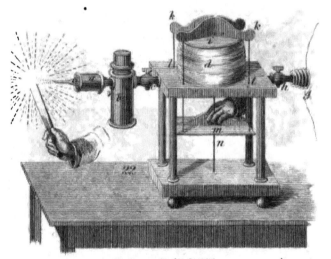

GLASGOW.
PUBLISHED BY R. GRIFFIN AND Cº

1827.
Engᵈ by H. Wilson.

A

PRACTICAL TREATISE

ON THE

USE OF THE BLOWPIPE,

IN

CHEMICAL AND MINERAL

Analysis:

INCLUDING

A SYSTEMATIC ARRANGEMENT OF SIMPLE MINERALS, ADAPTED TO
AID THE STUDENT IN HIS PROGRESS IN MINERALOGY,
BY FACILITATING THE DISCOVERY OF
THE NAMES OF SPECIES.

BY JOHN GRIFFIN,

AUTHOR OF CHEMICAL RECREATIONS.

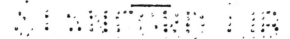

Glasgow:

PUBLISHED BY RICHARD GRIFFIN & CO.,

AND THOMAS TEGG, LONDON.

—

1827.

INTRODUCTION.

I. The little work, now offered to the public, is to be regarded in a two-fold point of view:—first, as a *Manual of Instruction* for the use of operators with the BLOWPIPE;—secondly, as an *Introduction* to the *Science*, and to the *Written Systems*, of MINERALOGY,—in other words, as a short and easy method of identifying mineral species.

That there existed a necessity for the publication of some such work as the present, I shall endeavour to show in the subsequent paragraphs of this Introduction. I cannot flatter myself with having produced exactly the thing which was required; but I have done the best I could to make the work a good one, and hope that, such as it is, it will be found useful.

While I trust that the inaccuracies to be found in it, are both few and unimportant, I may be permitted to observe, that the circumstances under which it has been composed, are such as entitle me to beg, that its faults may not be criticised over scrupulously. This is not, it must be remembered, the work of a professional man, not the gift of a master of science to its votaries; but, the offering of *a Student* to his fellows, the result of hasty researches made when his commercial avocations permitted him so to occupy a spare hour.

II. The BLOWPIPE is now in such esteem, both with artists and men of science, that little apology seems to be necessary for the publishing of a treatise, professing to exemplify its uses and to teach its manipulation;—more especially when it is considered, that a *general treatise* on this instrument, explaining the principles of its construction, the diversities of its form, and the many purposes to which it is applicable, does not exist in the English language.

So far as I am aware, there have been only two works published in this country, written expressly on the blowpipe; and both of these relate to *particular varieties* of the instrument. One of them, written by Dr Clarke, refers wholly to the gas blowpipe; the other, by Professor Berzelius, as exclusively regards the mouth blowpipe. The great bulk of the information which has been given to the world respecting the blowpipe, is contained in papers printed in various works upon other subjects; and is, therefore, not to be got at without considerable trouble and difficulty. The Essays of Bergman and Engeström, for example, were never published separately from the works in which they originally appeared, and which are now become scarce;—the instructions of Aikin are contained in his Manual of Mineralogy, a book that has long been out of print; —while the contributions of other writers towards a knowledge of the blowpipe, are disseminated through innumerable annals, encyclopædias, memoirs, and magazines. So that a person desirous of becoming acquainted with the general nature of this instrument, could only obtain a tolerable stock of information, by gathering together and by reading a great number of scarce, bulky, and expensive volumes, to the great waste both of time and of money.— Of course, this would not have been the case, had there existed a general treatise on the blowpipe. Here then was a desideratum. To supply this, by collecting the scattered intelligence I have spoken of, and condensing it into a useful form, was the object with which I set to work. The instructions comprised in the following pages are the result of the investigations undertaken with this view. How far my exertions have been crowned with success, I leave it for my readers to determine.

The various blowpipes which I have described, afford, I believe, specimens of every useful kind that has been invented. Many accounts of blowpipes which I met with, I passed over without notice, because they did not appear to me to present any thing novel or valuable. The instructions respecting the points of manipulation peculiar to each instrument, I have been particularly anxious to express clearly and distinctly, and hope I have succeeded in doing so. I have described all the fluxes and reagents commonly made use of, and exhibited the peculiarities of each of them. The tabular views of phenomena presented by minerals, when exposed under different circumstances to the action of the blowpipe, will be found in practice, to be extremely useful. In all cases, the facts stated in the early part of the work, receive ample illustration from the experiments recorded in the latter part.

III. I shall now make a few observations as to the uses of this work, considered as an Introduction to Mineralogy.

The first difficulty which the mineralogical student has to encounter, is that of identifying the various minerals which are presented to his notice. This difficulty, the wealthy student, amply furnished with specimens, and benefiting by the instructions of a living teacher, may experience only in a trifling degree ; but to him who is destitute of those aids,—to the solitary student, who, unaided by the advantages of personal instruction, or incapable of gaining access to a ticketed cabinet, depends for his success on his own industry and abilities, on the information derivable from books, and the examination of the few minerals which now and then may be thrown into his possession by chance; to the student thus situated, it is a difficulty of such mag-

nitude that comparatively few of those who begin
the study of mineralogy are ever able to master it.

Now, the object of the present work, is to furnish
the means of overcoming this difficulty; that is to
say, to point out a method whereby the mineralogical student may be easily enabled to discover the
name, and the place in a system, of any mineral he
may happen to meet with. I shall presently show
in what manner this is to be done; but it is necessary previously to explain, how it is that the difficulty alluded to originates.

When an individual, commencing the study of
mineralogy, has acquired the language of the science,
by making himself acquainted with the distinctive
characters of minerals, he is then able either to
describe a mineral himself, or to understand the
descriptions given by others. He may examine a
specimen, and write an account of its figure, structure, fracture, colour, lustre, transparency, and so
forth; and may point out the features wherein it
agrees with any other substance with which he may
choose to compare it. If he has learnt the name of
his specimen, he can refer to the " System" which
he may have adopted for his guide, and compare his
own description with that of his author. But if its
name is unknown to him, how is he to proceed then?
His " System" at once becomes useless; as the most
intimate acquaintance with the general characters of
minerals, can scarcely direct the student to the name,
or class, or genus, of a single species. So that he is
disabled from comparing his own description with
that of his author, and his progress is effectually put
a stop to. This disheartening circumstance arises
solely from our " Systems" or " Introductions" to
Mineralogy consisting merely of consecutive descriptions of minerals, unconnected by any easily-perceiv-
ed characteristics, and unaccompanied by any instructions whereby the names of minerals may be
come at. That amid the many works of this de-

scription which have been presented to the public, there should be none in which the classification is founded upon the properties or characters by which the substances treated of are distinguished, is to those acquainted with the kindred sciences of Botany and Zoology, a matter of no little surprise; because, in the works relating to their respective sciences, the very converse of this holds true. Thus, in Botany, for example, all vegetables are distributed into twenty-four Classes, distinguished by the number, proportion, or situation of the *stamina*. The classes are divided into Orders, taken from the number or differences of the *styles*, the *pericarp*, the number, proportion and situation of the *stamina*, or the disposition and character of the *florets*. The orders are sub-divided into *Genera*, containing all those which agree in their parts of *fructification*. The genera are again branched into *Species*, consisting of every individual whose form and structure is invariable and perpetual. —By the help of a system such as this, the student who examines a plant, plucked at random from the road side, is directed with readiness to a knowledge of its name. He then applies to his " System;" and learns the natural history of his specimen without the smallest waste of time. The success which attends his enquiries, urges him onwards in his pursuit, and he becomes enamoured of his study in proportion as he finds he is making progress.

How different is the case with the young mineralogist! He may pick up a mineral, as the botanist does a plant; but all he can do is to write a description of it himself:—unless somebody tells him the name of his specimen, he cannot refer to its natural history contained in mineralogical books. Let him turn to whatever " Introductions" he may, he will find in all the same want of that compendious mode of discovering the *names* of substances, unpossessed of which he never can make any progress in his pursuit, unless indeed he be possessed of a persever-

b 2

ance which is exhibited but rarely. Yet in these
" Introductions" to Mineralogy, there is no lack of
Systems, such as they are. The regular parapher-
nalia of science,—the classes, the orders, the genera,
and the families,—are to be found in every book;
but the mischief is, that these divisions and sub-
divisions are either made without any explanation,
are unaccompanied by characteristics, or are founded
upon the results of (often inaccurate) Chemical
analysis. So that, in a practical point of view, they
are utterly valueless, since they are not capable of
directing the inquiring learner to the object of which
he is in search. Their intricacies and contradictions
—whether they be mathematical or chemical classi-
fications,—are such, that common sense cannot un-
ravel them; and the clue which they are intended
to afford through the mazes and labyrinths of the
mineralogical domain, can neither be found nor
followed by any one who seeks for it.

Taking the subject of the classification of minerals
into serious consideration, we arrive at conclusions
something like the following:—There undoubtedly
ought to be a mineralogical system formed with the
utmost strictness of which science will admit; and
in which every known mineral should have its ap-
propriate place. If in the details of this system,
facility of reference could be associated with accuracy
of discrimination, the advantage would certainly be
great; but if it could not, convenience as a principle
should be disregarded. This is what is required by
philosophical precision, in a System or Classification
intended to form the basis of the science. But, at
the same time, we are told by experience, that, with
this scientific system, there should be an artificial
arrangement, constructed solely with regard to its
conveniency; intended, as Berzelius says, " to rank
after the proper system, as an index ranks after a
book; and adapted for those who study mineralogy
without the assistance of an experienced master and

an ample collection, and who are often obliged to enquire the names of minerals with which they are unacquainted."—For the want of an arrangement of this kind, capable of enabling the unassisted student to identify mineral species, the works of the most eminent professors, are, so far as he is concerned, in a great measure unintelligible and useless;—as the examination of a single system would be sufficient to demonstrate.

But it is unnecessary to occupy the reader's time by critically attempting to prove a fact, of which, if he be a mineralogical student, he will probably be already but too well aware; and indeed, it would be invidious to point out any particular work, as defective and censureable in the respect alluded to. It is sufficient for the present purpose to remark, that although every mineralogical " System" contains an attempt at the formation of a pure philosophical classification, scarcely a single work can be found in which the author has condescended to provide, for the convenience of students, an arrangement upon an artificial basis. The effects of this may be stated in a few words:—When a student is desirous of knowing the natural history of a mineral of which he has obtained a specimen but not learned the name, his only plan of procedure is to toil through the volumes of his mineralogical · guide, examining, as he proceeds, the characters of every species that meets his eye. He who has never thus toiled through the *Eighteen Hundred Octavo pages* of which Professor Jameson's System is composed, can form no idea of the drudgery of the business. Yet, this drudgery there is no avoiding : the student must go through it. Perchance his labour may be rewarded by the discovery of the species he is in search of; but the greater probability is, that, after floundering on a long time, in doubt and uncertainty, he will at length become completely bewildered in comparing the characters he meets on every

page with those which distinguish his specimen;
and eventually will be obliged to give up the task
he has engaged in, in utter despair.—If this is not
enough to damp the spirits of a student, to put an
effectual stop to his progress, and to make him turn
from the pursuit in disgust, I do not know what is.
When we call to mind the many difficulties of this
sort which beset the student of mineralogy, the
wonder is, not that so few persons learn the science,
but that it is ever learned by any one at all.

The only mineralogical works in which any attempt has been made to supply the deficiency I have
pointed out, are Aikin's Manual and Bakewell's
Introduction. The first of these works contains a
synoptical table in which the principal minerals described in the book, are classed in groups, chiefly
with a relation to their habitudes before the blowpipe. The other work contains a tabular view of
the most important earthy minerals, arranged according to their hardness and specific gravity.

The Systematic Arrangement presented in the
following work, is founded upon a consolidation of
the two plans above referred to. The outline thus
obtained has been filled up by information drawn
from a great variety of sources, I have carefully
inserted in its proper place every mineral which the
student is likely to meet with, and have described
its most important and characteristic properties. I
have also been particular in giving the *Synonymes* of
the minerals, that is to say, the names by which they
are recognised by various authors; so that this will
be a useful adjunct to almost any English work on
the science. The great assistance towards the construction of an artificial system of mineralogy, derived
from the numerous and accurate blowpipe experiments made of late by Berzelius, has enabled me to
remove from the common-place descriptions of minerals a number of very important errors; and I
trust, has contributed to render the synopsis which I

presume to offer to the public, more correct, and better adapted to answer its intended purpose, than any thing of the kind that has yet been constructed. I by no means intend to insinuate, however, that the work is free from inaccuracies. It would be a wonder, indeed, if, amid so many thousands of facts a number of fictions were not lurking. But I hope that no greater portion of errors will be detected, than might be expected to occur in a work necessarily attended with many difficulties, and of which the writer is peculiarly liable to be deceived by erroneous statements.

I wish it to be clearly understood by the student, that the Systematic Arrangement here presented is merely intended to enable him to discover the *Names* by which minerals are known to the writers of Mineralogical Systems, which is the only thing that is wanting to render those systems intelligible and useful. It is by no means intended that this work should supply the place of a regular System:—on the contrary, it is designed to be an introduction and companion thereto.—But, at the same time, since it contains a comprehensive summary of the most important characters of minerals, it may occasionally, especially by a traveling student, be made to supply the place of a more bulky production.

When the young mineralogist endeavours, with the aid of the following work, to learn the name of a mineral, his first business is to determine to which of the Classes it belongs. The Essential Characters of the Classes, described at paragraphs 389—392, will enable him to do this with little difficulty or risk of error,—the greatest attention having been bestowed upon these characters to render them not only precise and accurate, but easy to be understood. When the Class of a mineral is determined, a very considerable progress is made towards the attainment of the ultimate object. Were there no other divisions, the student would have merely to compare his mineral with those comprised in the same Class, instead of having to run through all the minerals in the System. But there are farther divisions, and divisions, too, of which the characters are much more perfect than are the characters of the Classes. But to return. The Class being determined, the Order is next to be sought. If the mineral belongs to the Combustible Class, its Order may be discovered by the characters described at paragraph 397; if to the Metallic Class, by those described at 414; if to the Earthy Class, by those at 548; if to the Saline Class, by those at 739. As was before mentioned, when the Class is determined, the determination of the Order will be effected with but little difficulty; and when the Order is determined, the discrimination of the Genera and Families, where there are any, will be still less difficult. In this Classification, every successive division is founded on characters that are peculiar to the bodies which constitute it; and these peculiar or distinctive characters, which constitute the basis of the arrangement, are placed at the beginning of the divisions that they relate to. Hence, the enquirer, examining them as he proceeds, is directed straight onwards to the object he is in search of, like the traveler who meets at every corner with signs to direct his course.

It is unnecessary to occupy the reader's time, by extending these general observations. I shall only add, that I hope this humble attempt to smooth the path to Mineralogy will be received with candour, and that it will be judged of with due allowance for the manifold difficulties by which the subject is attended.

Glasgow, Sept. 1st, 1827.

TABLE OF CONTENTS.

---

c

PART II.

A SUMMARY VIEW OF THE SIMPLE BODIES, AND PRINCIPAL CHEMICAL COMPOUNDS, WHICH ENTER INTO THE COMPOSITION OF MINERALS, AND OF THE PHENOMENA WHICH THEY PRESENT WHEN EXPOSED TO THE ACTION OF THE BLOWPIPE.

PART III.

A SYSTEMATIC ARRANGEMENT OF SIMPLE MINERALS.

CLASS 4. SALINE MINERALS.

PART IV.

MISCELLANEA.

PLATE 1.

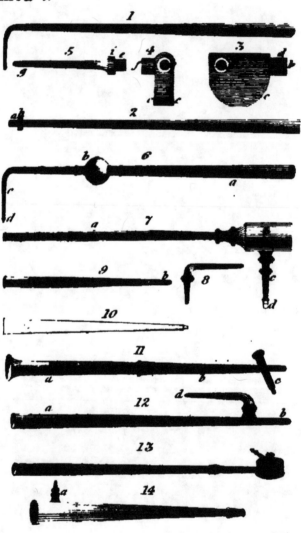

H. Wilson

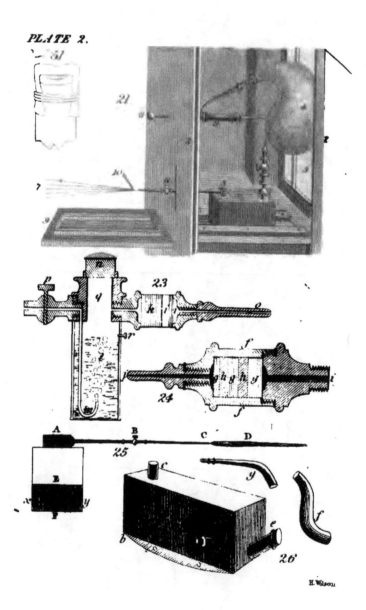

PLATE 2.

H. Wilson

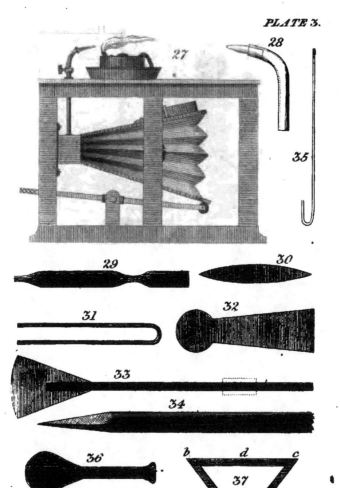

PLATE 3.

J Wilson

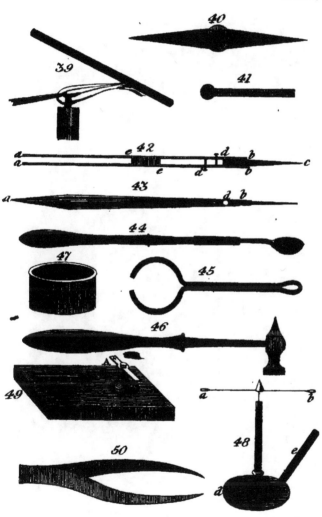

PLATE 4.

A
PRACTICAL TREATISE
ON THE USE OF
THE BLOWPIPE.

HISTORY OF THE BLOWPIPE.

1. THE BLOWPIPE is an instrument by means of which the flame of a candle or lamp, can easily be converted into a species of furnace, capable of communicating an intense heat to small bodies placed in the flame, and consequently of being applied to many useful purposes. The operation consists in impelling, from a small aperture, against the flame, a stream of air moved with velocity, by means of the organs of respiration and the mouth, or by a bellows. The blowpipe is used in soldering, by the jeweller and goldsmith, and other artists who fabricate small objects of metal; by the glass-blower, in making thermometers, barometers, and other instruments and toys from glass; by the enameller; and in glass-pinching, which is the art of forming glass in a mould fixed on a pair of pincers, into the ornamental pendants for glass-lustres. This latter is one of the many ingenious processes carried on at Birmingham. It is also employed by a variety of other artists, as a convenient and powerful assistant.

2. But, notwithstanding that this instrument was employed in the arts at a very remote period, it is only about a century since any one conceived the idea of applying it to experiments in chemistry.

A

The first philosopher who used this elegant apparatus for pyrognostic operations was Andreas Swab, a Swedish Metallurgist, and Counsellor of the College of Mines, about the year 1738; at least, such is the information given to us by Bergman. Swab, however, left no work on the subject, and it is unknown to what extent his researches with this instrument were carried. We have therefore to depend on later experimenters for the knowledge which is to direct our inexperience.

3. After Swab, the first person of eminence who used the blowpipe was Cronstedt, the founder of Mineralogy; a man whose genius so outstripped the age in which he lived, that his contemporaries found his writings unintelligible. The purpose for which he employed it, was the discrimination of minerals by means of fusible reagents. His experiments were conducted with a degree of perfection that could only have resulted from the most persevering industry. He published in 1758 such results of his operations as the nature of his mineralogical work required; and it is to be lamented that the circumstances of the times in which he lived, were such as to prevent his making fully known to the world this new application of the blowpipe, and describing in detail the processes which he had adopted so successfully.

4. In 1770, appeared an English translation of Cronstedt's work, by Von Engeström, annexed to which was a treatise on the blowpipe, wherein many of Cronstedt's processes and their principal results, were described very accurately. This production had the effect of drawing the attention of chemists and mineralogists to the use of this instrument: it made however few adepts; for, in matters of practical Science, books alone are but "weak masters." In Sweden only did the pyrognostic art make any considerable progress; *there*, those who had seen Cronstedt and Von Engeström at work, learned to work like them, and transmitted their skill to their successors.

5. The great Bergman soon outstripped all his predecessors. The blowpipe in his hands became one of the most valuable instruments of analytical research. He extended its use beyond the bounds of mineralogy, to the field of inorganic chemistry;—applied it to such a diversity of purposes; conducted all his operations with such tact; and brought out his results with such nicety; as at once to demonstrate the vast advantages capable of being derived from the use of the blowpipe, and his own unrivalled skill as an analytical chemist. His work De Tubo Ferruminatorio, may be regarded as a masterpiece of philosophical investigation,—equally simple and beautiful, accurate and profound. It was published at Vienna, in 1779, and Dr Cullen's translation of it, in London, in 1788. We subjoin from this work, a brief enumeration of the advantages of the blowpipe:—

6. The Blowpipe is an instrument of extreme utility to chemists—since many experiments are daily neglected, either because they require large furnaces and cumbrous apparatus; or, from the want of the time necessary to conduct them in the ordinary way; or from the difficulty of obtaining the quantity of matter required for an experiment in the common way, when it happens to be very scarce or very dear. —In all these cases, the blowpipe may be used advantageously; as, 1. Most of the experiments which can be performed in the large way, may also be done with the blowpipe. 2. The experiments, which in the large way require many hours, may in this method be finished in a few minutes. 3. The smallest particle of a substance is sufficient.

7. These conveniences, however, though of great weight and importance, are attended with this defect, that they do not determine the proportions of the ingredients indicated. For this reason, when experiments can be performed on a large scale, that plan is to be adopted as the most accurate. But the first inquiry to be made, is, generally, *what* a sub-

stance contains, not *how much;* and the experience
of many years has taught me, that these trials in
small quantities suggest the proper methods of in-
stituting experiments in the large way. Besides,
these experiments have some exclusive advantages
over those performed in crucibles.—1. We can see
all the phenomena which occur from the beginning
of a process to the end; and thereby gain information which wonderfully illustrates the series of operations and their causes. 2. Experiments made in
crucibles, give often fallacious results, owing to the
corrosion of the substance of the vessel: we suppose, for instance, that lime or magnesia melted
with fixed alkali, are united with it in the way of
solution; but the globule, when well fused in the
blowpipe spoon, permits us, by its transparency,
clearly to perceive, that, except the siliceous part, it
is only mechanically mixed. 3. The most intense
degree of heat may in this way be obtained in a few
minutes, which can scarcely be done in a crucible
in many hours.—It is needless to multiply commen-
dations of the blowpipe: its valuable properties will
be rapidly developed to those by whom it is employed.—Such were the opinions of this philosopher
with regard to the utility of this little instrument.

8. The continual application which Bergman bestowed on his studies, had so great an effect on his
health as to oblige him to continue his philosophical
pursuits with the aid of an assistant. He accordingly employed Assessor Gahn, who performed under his directions a series of operations on all the
minerals then known, by which he was taught in
what manner each individual conducted itself before
the blowpipe. The experience thus acquired, enabled Gahn to employ the instrument in every kind
of chemical and mineralogical enquiry; by this
means he attained so great a degree of skill, as to be
able to detect the presence of substances which had
escaped the most careful analysis. Berzelius informs

us, that, long before the question was started, whether the ashes of vegetables contain copper, he had seen Gahn many times extract, with the blowpipe, from a quarter of a sheet of burnt paper, distinct particles of metallic copper. As Gahn was always accompanied by his blowpipe on his travels, and made continual use of it, he was led to contrive numerous improvements in the manner of using it; and his manipulations were imagined and executed with so much sagacity and precision, that his results were entitled to the greatest confidence. He seems never to have thought of publishing an account of his labours; although he was always ready to instruct those who wished for information on the subject, and although by doing so he would have conferred upon the scientific world an undoubted obligation.

9. The eleventh volume of the Annals of Philosophy, contains an article "On the Blowpipe; from a Treatise on the Blowpipe, by Assessor Gahn, of Fahlun." The greater part of this article is copied into Mr Children's Translation of the Fourth Volume of Thenard's Chemistry, and into Dr Ure's Chemical Dictionary. These are the only sources, at least in the English language, whence any information can be derived respecting the services which Gahn's blowpipe has rendered to science.

10. Berzelius was, for the last ten years of Gahn's life, his pupil and friend, and during the familiar intercourse of that period, gained from the latter a considerable portion of that knowledge, which was the result of the experience of a long and active life. A short time previous to Gahn's death, Berzelius undertook an investigation, the object of which was to ascertain, and to note down in a systematic manner, the phenomena presented by different minerals when acted on by the blowpipe. The results of this, Gahn, with his blowpipe in his hand, was to have examined, criticised, and corrected; in order, if possible, to get rid of every in-

accuracy. This fair project was frustrated by Gahn's
unexpected death.

11. But, notwithstanding this disappointment, Berzelius proceeded with the course of experiments
which he had instituted; purposing to supply the
loss of the aid promised by his friend and master,
by a double portion of his own industry and sagacity. That he succeeded most completely in the
labour which he had taken in hand, the treatise he
has published on the blowpipe is a sufficient demonstration. He has finally established the fact, that
the blowpipe is, to the mineralogist, an instrument
of *indispensable utility;* and the volume wherein his
experiments are detailed, is one of the most useful
books on practical chemistry extant. The English
mineralogist is much indebted to Mr Children, by
whom a translation of Berzelius's work was published in 1822.

12. All that has hitherto been said, refers to that
description of blowpipe, whatever its form may be,
through which air is impelled from the mouth. But
when the blowpipe deservedly came to be considered as an essential instrument in a chemical laboratory, several attempts were made to facilitate its use,
by the addition of bellows or other equivalent instruments, adapted to continue the supply of air,
and at the same time to leave the operator at liberty. Though these render it less portable, and consequently less useful for mineralogical researches,
some of them are unquestionably well adapted for particular purposes: to pass them over, without notice?
would therefore be wrong.—Berzelius, it is true, con-
demns these contrivances. "By these pretended
improvements," says he, "motions, more or less
troublesome, have been substituted for a slight exertion of the muscles of the cheeks; and their inventors
have demonstrated, by their very contrivances, that
they did not know how to use the blowpipe: they
might as well have proposed to play on a wind in-

strument with a bladder. Our conclusion must be, that all apparatus of this kind is perfectly useless." It is to be presumed, from this passage, that Berzelius is blessed with very good lungs; and it is to be lamented that every one else has not as good as he. Doubtless, for general purposes, the mouth blowpipe is, without exception, the best in existence;—nevertheless, there are many uses for which its power is quite inefficient, and for which we require more effectual aid. Certainly, then, we ought not to scorn what we are sometimes obliged to employ.

13. Besides those compound blowpipes, the use of which depends upon mechanical aid, others have been contrived which rest their claims to pre-eminence upon the advantages gained by employing *gases* instead of common air. These are deserving of attentive consideration. They have higher aims than the aforementioned machines, and in several cases have been eminently successful. No sooner was *oxygen gas* discovered, than Lavoisier applied it to combustion in the manner, and with the success, stated in the following passage, copied from the English translation of his Elements of Chemistry:—"I endeavoured to employ oxygen gas for combustion, by filling large bladders with it, and making it pass through a tube capable of being shut by a stop-cock; and in this way I caused it to support the combustion of lighted charcoal. The intensity of the heat produced, even in my first attempt, was so great as readily to melt a small quantity of crude platinum. To the success of this attempt is owing the idea of the *gasometer*, which I substituted instead of the bladders; and, as we can give the oxygen gas any necessary degree of pressure, we can with this instrument keep up a considerable stream, and give it even a very considerable force." In a succeeding part of this work are given the results of some experiments made by Lavoisier with this blowpipe, in 1782.

14. But oxygen is not the only gas capable of being used with the blowpipe; as even gases which are said not to be supporters of combustion can be applied to that purpose. It is a curious fact, that pure hydrogen gas, when highly compressed, and propelled through a capillary tube, exhibits, during combustion, a very exalted temperature. We are informed by Dr Clarke, that he succeeded in fusing platinum foil by its means, and that the combustion of iron wire by burning pure hydrogen gas in this manner, is an experiment always attended with success!

15. Immediately after the discovery of the composition of water, the combustion of oxygen and hydrogen gases together were successfully applied to the purpose of exciting an intense degree of heat by the blowpipe. The peculiar construction of the apparatus cannot be understood without a plate, which we have not thought it worth while to have engraved for this work, but which may be seen in the *Annales de Chimie*, tom. xlv., or in the 14th volume of the Philosophical Magazine. It is sufficient to observe here, that the gases were expelled from separate gas-holders, and that, in order to provide against explosions, their mixture did not take place, till they nearly reached the aperture of the pipe, at the extremity of which they were inflamed. We give but a brief description of this contrivance, although it was one of considerable importance, because the employment of it was completely set aside by the greatly-superior invention of the *Gas Blowpipe* of Dr E. Clarke; of the circumstances connected with the discovery of which, we are now about to speak.

16. Sir Humphry Davy having declared that the explosion from mixed oxygen and hydrogen would not communicate through very small apertures, Mr Children proposed to him to employ Newman's blowpipe for effecting the combustion of a condensed mixture of oxygen and hydrogen gases, issuing from

a small aperture. This Sir Humphry did, and found that the flame produced a most intense heat, which instantly fused bodies of a very refractory nature. Dr Clarke, Professor of Mineralogy at Cambridge University, having consulted Sir Humphry on the subject, proceeded to expose a great variety of mineral substances to the flame, for the purpose of observing its effects upon each of them. The tube through which the mixture of the two gases issued, was cemented on the pipe of issue of Newman's blowpipe. The tube at first used by Dr Clarke was three inches in length, and the diameter of its cavity the seventieth part of an inch. The end of this tube was constantly breaking during the experiments, owing to the sudden changes of temperature, until at last he usually worked with a tube only one inch and three-eighths in length. When the current of gas was feeble, from the gas in the reservoir having become nearly of the same degree of density as the surrounding air, or from the current being suppressed in the beginning of an experiment, the flame then had a retrograde motion, passed about half an inch up the capillary cavity of the tube, exploded, split the end of the tube into pieces, and then was extinguished, without doing any farther damage. In order to try the effects of an explosion, four pints of a mixture of the two gases were condensed into the chest, that quantity being all the syringe would force into it ; and, the glass tube being taken off, so that the diameter of the nose-pipe, by which the gas was to issue, was about one-eighth of an inch ; a burning spirit lamp was placed at the aperture ; when, the stop-cock being opened by means of a long string attached to it, the whole gas exploded with a report like that of a cannon : the chest, formed of strong plates of copper, was burst asunder, the stop-cock was driven out, and one end of the chest was torn off, and violently thrown against the wall of the room. This showed the danger of using the apparatus with too large an aperture, and the necessity of employing a capillary tube.

17. It was not long after these experiments before Dr Clarke made the capital discovery, that to produce the greatest effect with this blowpipe, it was necessary to mix the gases in the exact proportion in which they combine to form water, namely, two volumes or measures of hydrogen to one of oxygen gas. The gases being obtained in a state of great purity, and mixed in the proportions here stated, occasioned a heat greater than that produced by the largest galvanic batteries. A great number of substances ever before deemed infusible, were completely melted by it; and many of the phenomena produced were equally curious and astonishing.

18. " But," says Mr Children, speaking of the effects of Dr Clarke's oxy-hydrogen blowpipe, "one essential character, the fusibility, or infusibility, of different substances, as determined by the common blowpipe disappears before the intense heat produduced by this, *which levels all bodies to one general mass of fusible substances;* though very evident differences are still observable in the *facility* with which different bodies are reduced to the state of fusion. In return, too, for the character which is thus lost, we gain a new one in the appearance of the, otherwise infusible body, after it has been melted."

19. Numerous accidents which occurred to Dr Clarke and others, by the explosion of the reservoir, some of which had nearly been attended with serious consequences, occasioned several attempts to be made at improving the apparatus. The most perfect of these produced for some time, with respect to *safety*, was the trough suggested by Mr Professor Cumming of Cambridge; a description of which is given in a subsequent part of this work. With this improvement, Dr Clarke used the gas blowpipe at his daily lectures during a whole session, without the occurrence of any accident whatever; establish-

ing thereby its safety and utility beyond all question.

20. No farther alteration was made in the gas blowpipe till 1823, when Mr Goldsworthy Gurney, of the Surrey Institution, published an account of a new instrument invented by himself. This blowpipe was so constructed, as to enable the operator to produce a flame of great size, power, and brilliancy; by burning *large quantities* of the mixed gases, with the *utmost safety*. A complete description of this apparatus being given elsewhere in this volume, it is unnecessary to enter into details here.

21. As no invention or discovery of any importance relating to the blowpipe, has been made of late, it is unnecessary to extend farther the bounds of its history, which we therefore now bring to a close.

DESCRIPTION OF BLOWPIPES.

22. Blowpipes are of two kinds, *simple* and *compound.*—1. *SIMPLE blowpipes* are those which consist merely of a tube through which air is blown from the mouth.—2. *COMPOUND blowpipes* consist primarily of a tube through which air is blown from some description of secondary apparatus attached to it. Of each of these two sorts of blowpipes, many different inventions have at various times been recommended to public notice; some of them valuable from their simplicity, cheapness, power, or convenience; others possessing a kind of merit which may be imagined from our describing the machines as exceedingly ingenious, expensive, and useless. From the great mass of these inventions we have selected those whose good qualities are perceptible to other persons than the inventors; and of these we shall subjoin such figures and descriptions, as we

hope will enable the student to comprehend their
construction and relative advantages.

1.—OF SIMPLE BLOWPIPES.

23. *It is necessary to premise, that, whatever may
be the* FORM *of a blowpipe, the* LENGTH *of it must
depend on the eye of the operator. The body operated upon ought to be at that distance from his eye at
which he has the most distinct vision: of course,
therefore, the length of his blowpipe must be regulated
by the strength or weakness of his sight. The most
common lengths, however, are from* SIX *to* EIGHT
inches.

24. With respect to the *MATERIAL of which blowpipes should be made,* there is a great diversity of
opinion. It must be of such a nature as to be easily
wrought into the form required, and at the same
time ought to be durable and cheap.—Berzelius says,
" blowpipes are best made of silver, or tinned iron
plate, the beaks only being of brass. If the instrument be wholly of brass, it in time acquires the
taste and odour of verdigris,—an inconvenience not
entirely removed by making the mouth-piece of
ivory. The hands, too, if not quite dry, contract
the same odour during the operation, especially if
the blowpipe has not been used for some time, and
was not well cleaned before it was laid by. Tin
plate is not liable to this nuisance, and has, besides,
the advantage of cheapness. When that material is
employed, the joinings should be made air tight by
inserting pieces of paper between them. Notwithstanding that silver is the best conductor of heat of
all the metals, no inconvenience need be apprehend-
ed on that score, even in the longest operations.

25. " The small jets adapted to the extremity of the
beaks are a great improvement, for the extremity is
liable soon to become covered with soot, and the
hole to be either blocked up, or lose its circular

form ; and it is necessary to clean it, and clear the opening with a small needle kept at hand for the purpose. This is an indispensable but troublesome operation. I have, therefore, had these jets made of platinum, each of a single piece, and when dirty I heat them red hot on a piece of charcoal, which cleans them in an instant, and clears the hole without the assistance of any mechanical agent. Silver would not answer the purpose, for although we were ever so careful not to fuse it in the operation, it would crystallize as it cooled after having been heated to redness, and become as brittle as an unmalleable metal.

26. "Glass blowpipes are certainly less costly, and less liable to get dirty than those made of metal; but their brittleness and the fusibility of their beaks are so serious inconveniences, that they should never be used but in cases of necessity."

27. Children remarks upon these observations, that in this country the blowpipes are generally made of brass, and, when well lackered, do not annoy the operator with the inconveniences of which Berzelius complains. The mouth-piece should be of silver, and the beak of platinum; were it not for the expense, the whole instrument were better made of platinum.

28. Glass blowpipes have these advantages.—1. It is easy to give them the requisite form, and to make the aperture smooth and round. 2. The material is cheap. 3. They can easily be made from glass tubes by any one who can use the blowpipe at all. To the operator who lives in a retired situation, where artists are not easy to be come at, these advantages recommend glass strongly.

29. *The Common Blowpipe.*—In its best-known and most simple form, this instrument is a conical tube or pipe, of brass or white iron, about eight inches long, and one-fourth of an inch in diameter at the top ; with a curvature near the lower end,

B

whence it tapers off to a point, which has a very
small perforation for the wind to escape. Fig. 1,
will render this description perfectly intelligible.
This is the original form of the blowpipe, the shape
which was adopted by the artists of old; and it is
worthy of observation, that, notwithstanding the
innumerable alterations and improvements of it,
which philosophers have thought expedient, the
working goldsmiths and other artizans who chiefly
employ the blowpipe, still prefer the old shape, and
by doing so, unquestionably show their good sense:
it is the cheapest, the simplest, and the handiest;
and for short and common jobs, excels all others.

30. *Cronstedt's Blowpipe.*—When Cronstedt began to use the common blowpipe for chemical experiments, he was greatly annoyed by the condensation
within the tube, of the vapour of the breath, which
either impeded the blast, or was projected into the
flame. To remedy this inconvenience, he placed
towards the small end of the instrument, a hollow
ball, intended to collect the condensed vapour. This
was of considerable service, but, did not effect a ra-
dical cure of the evil; for, when the beak of the
pipe happened to be held downwards, the water still
ran out. Bergman, into whose hands this blowpipe
came, found it necessary to get rid of defects which
were a source of much inconvenience; this led to
another improvement.

31. *Bergman's Description of his Blowpipe.*—That
which I use, (says he) is formed of pure silver, lest
it should be injured by rust; a small addition of
platinum communicates the necessary hardness. It
consists of three parts. A handle, fig. 2., termina-
ting in a truncated conical apex *a*, which may be,
by twisting, so adapted to the aperture *b*, fig. 3, as
to shut it more closely than can be done by a screw.
To collect the condensed moisture of the breath, I
make use of the little box, fig. 3, 4, formed of an
elliptical plate, so bent at the centre that the oppo-

site sides become parallel, and are joined together
by a plate, as represented at c, fig. 3, and c o, fig. 4:
such a box has an advantage over a spherical one,
inasmuch as it occupies less space, and is more convenient in use. The aperture b, which is somewhat
conical, and hollowed out of the solid piece, has no
margin turned inwards, lest the efflux of the fluid,
collected after long blowing, or the cleansing of the
internal parts, should be in any degree prevented.
The tube, fig. 5, is very small, and its shorter conical end e exactly fitted to the aperture f, so that no
air can escape, unless through the orifice g. Many
of these tubes should be provided, with orifices of different diameters, to be used on different occasions.
The orifice g must be truly circular and very smooth,
otherwise the flame which will be produced by it, will
not be of the requisite shape. The bands, h on fig.
2, and i on fig. 5, prevent the conical apices a and e
from being thrust too far into their receptacles; they
also serve another purpose : when the apices, by repeated attrition, become too small to fit the cavities
intended to hold them, they are again rendered tight
by filing away a little of the bands.

32. *Magellan's Blowpipe.*—Fig. 6, is a representation of Cronstedt's blowpipe, with an improvement
of which a description is given in Magellan's edition
of Cronstedt's Mineralogy, published in 1788. For
this improvement, Mr Children believes we are indebted to Mr Pepys: it may be so; but, from the
above circumstance, I am inclined to think otherwise. a and c are the two pieces that form the blowpipe. The ball b, is hollow, and formed of two he-
mispheres which screw together: the male screw is
in the lower hemisphere, which is soldered on the
part c at such a distance, that the inside end of the
crooked tube is even with the edge of the hemisphere, as represented by the dotted lines. The upper hemisphere is soldered at the end of the straight
tube a. By these means, the moisture formed by

the breath, falls into the hollow of the lower hemisphere, where it is collected round the upper inside end of the bent part *c*, without being apt to fall into it.

33. The small nozzles, or hollow conical tubes, advised by Engeström, Bergman, and others, says Magellan, are wrong in the principle ; because the wind that passes from the mouth through such long cones, loses its velocity, by the latteral friction, as happens in hydraulic spouts, which, when formed in this manner, never throw the fluid so far, as when the fluid passes through a hole of the same diameter made in a thin plate of a little metallic cap that screws at the end of the large pipe. It is on this account that the little cap *c* is employed, having a small hole in the thin plate, which serves as a cover to it. There are several of these little caps, with holes of smaller and larger sizes, to be changed and applied, whenever a flame is required to be more or less strong.

34. Another convenience of these little caps is, that even in case any moisture should escape falling into the hemisphere *b*, and pass with the wind through the crooked pipe *c*, it can never arrive at, nor obstruct, the little hole of the cap *d*, there being room enough under the hole in the inside, where this moisture is stopped, to retain it till it is cleaned and wiped out.

35. Whatever merits may be possessed by the other blowpipes described in this work, it is certain that this one has remained till the present time a very general public favourite.

36. *Gahn's Blowpipe.*—Although Bergman's improvement of the blowpipe effected the object he had in view, yet Gahn found the instrument capable of being still further improved. He gave to his receiver the form of a cylinder, certainly a more convenient shape than Bergman's—the dimensions of which were, one inch in length and half an inch

in diameter. In other respects his instrument was on the plan of Bergman's, and consisted of four pieces, a, b, c, d, fig. 7. The little jet, d, is fitted by grinding to the extremity of the beak c : there are several of these jets with holes of different diameters, to be changed as occasion requires. For some purposes, such as the blowing of glass, and other operations, the bent beak, fig. 8, is preferable to the straight one c, fig. 7. Berzelius declares this blowpipe to be preferable to all others.

37. *Tennant's Blowpipe.*—This, which is represented by fig. 12, lays claim to public favour on the ground of its portability. It is a straight tube, slightly conical, closed at its narrow end b, at half an inch from which is an opening to receive the small bent tube d, which is fitted in by grinding, and can be turned in any direction required. As the water condensed from the breath flows to the point b and does no harm ; and as the instrument is very portable, very simple, and capable of affording a powerful blast; it appears to combine almost every advantage that can be wished for in a blowpipe.

38. *Wollaston's Blowpipe*—is still more portable than the preceding. Fig. 9, 10, 11, show its form, the number of pieces of which it is composed, and the manner of putting the parts together, whether for use or for traveling. a, b, c, fig. 11, represent the three pieces put together for use. The point of b is closed, but a little way from it, on the side of the tube, there is a small opening, as seen at b, fig. 9, which emits the air into the beak c, fig. 11, when that is brought into its proper situation on the tube. Fig. 10, represents this blowpipe, which exceeds all others for portability, packed up into the space of a toothpick or pencil case, for traveling. If we add to this, a slip of platinum foil, two or three inches long, to hold the object of experiment to the flame, and a small piece of borax to serve as a flux, we are furnished at once with a sufficient laboratory for a

great variety of experiments ;—for the candle and
charcoal may be found in most places. He who may
deem this docimastic apparatus not sufficiently commodious, can enlarge it at pleasure, by adding to it
any of the instruments described hereafter.

39. *Pepys' Blowpipe.*—In this, the contrivance for
containing the condensed vapour, is a flat cylindrical
box, which is also made to answer the purpose of
holding additional caps for the nozzle. Fig. 13, is
a representation of this blowpipe, which in Dr Hen-
ry's Elements of Chemistry, is designated as " a
most commodious" one.

40. *Dr Black's Blowpipe.*—Figure 14, is of tin,
that is to say, of tinned iron ; the small pipe *a*, is of
brass, and has two or three caps that fit on tight ;
each cap is pierced with a hole of a different diameter, and consequently the force of the blast may
be varied at pleasure, by changing the cap.

2. OF COMPOUND BLOWPIPES.

41. Having given a description of the best simple
blowpipes, we shall now do the same for a few of the
multifarious machines which may be termed compound blowpipes.

42. *The Fixed Blowpipe.*—" De Saussure," says
Berzelius, "fixed his blowpipe to a table, so as to
have his hands at liberty, while he regulated the
blast by his mouth. I know no case in which
much advantage is derived from this plan." Mr
Children subscribes to the opinion that these contrivances are useless, and adds, " the skilful workman
needs no such aids, and the operator with the blowpipe will do well to render himself independent of
them at once. However, as in a few cases, where an
unusually large flame is required, a fixed instrument
may be useful, I annex a figure of the best form
I have met with, for a blowpipe on that construction. I do not know who is the author of the inven-

tion. *a* fig. 26, is a rectangular copper box, 2½ inches long, one inch high, and five-eighths of an inch wide; which is fixed on a board by screws passing through a foot plate *b*. *c* is a tube projecting from the top of the box, to which one end of a flexible tube may be adapted by a brass socket. The flexible tube is terminated at the other end by an ivory mouth-piece, and is of such a length as to be conveniently applied to the mouth of the operator. *d* is another projecting tube to which the beak *f g*, may be adapted; and *e* is a third projecting tube (closed with a cork) for the purpose of pouring off the water condensed in the box, *a*. By twisting the different parts of the beak in different directions, the jet may be presented to the flame in any position which may be required. All the parts that have motion are well fitted together by grinding. *f* and *g* may be made either of metal or glass."

43. *Hooke's Self-acting Blowpipe.*—M. Pictet conceived the ingenious idea of making the eolipile serve the purpose of a blowpipe. The effect was intended to be produced by the vapour of boiling alcohol. Mr Hooke contrived the apparatus represented by figure 20. A, a hollow copper sphere, two or three inches in diameter, for containing highly-rectified alcohol; this sphere rests on a shoulder in the ring *g*. A bent tube, *b*, with a jet at the end is screwed into the sphere for conveying alcohol in the gaseous state to the flame at *o*. This tube is continued within the sphere almost to the top, so that the vessel may be nearly filled with alcohol. At the top of the sphere there is a safety valve, *c*, to prevent those accidents, which might otherwise arise from the expansion of the fluid, in case the heat applied were too great. The pressure of this valve may be regulated at pleasure, by the two milled nuts, *e, f,* and the steel arm *h*, which presses on the valve *c*. The lamp which heats the alcohol is marked *k*. It is made to adjust to different distances from the sphere,

A, by sliding up or down between the two pillars, *l*, *l*. The distance of the flame *o*, from the jet of the tube, is regulated by the piece *m*, which by means of a screw, gives the wick-holder of the lamp an eccentric motion, so as to carry it from the centre. The opening for filling the globe with alcohol is marked *i*: it is secured by means of a milled finger-button, and a collar of leather. *n* is a mahogany stand supporting the whole. *z* is a jet of the vapour of alcohol inflamed by the lamp. It is needless to do more than merely describe this blowpipe, since in reality it is much more ingenious than useful. It was proposed to apply it to the purposes of the mineralogist; but it does not appear to be either so readily put in action, or so efficacious as the common blowpipe; which is also simpler in its construction, less bulky, and more easily carried about. Moreover, the flame of the vapour of alcohol, is far less steady and equable than that urged by a current of air; so that upon the whole the spirit blowpipe can only be regarded as a pretty philosophical toy.

44. Mr Nicholson, in the ninth number of his Journal, mentions a little implement, or eolipile, sold by the glass-blowers, and acting on the same principle. It consists of an egg-shaped ball, having a stem or handle, and a nozzle or tube. When half-filled with water or spirit, a strong current is produced, with which, blown through the large flame of a tallow lamp, Mr N. fused a half-penny, and softened the glass stopper of a bottle, so as to run a wire through it without bursting the vessel.

45. *Glass-Blowers' Blowpipe.*—When it is required to continue the use of the blowpipe so long as would be fatiguing if the breath merely were employed, the glass-blowers' table is employed. It consists of a double bellows, so fixed as to be worked by the foot, and to impel a current of air through a tin blowpipe, against the flame of a lamp fixed on the table. For the sake of durability, the blowpipe is sometimes

made of brass on which is screwed a nozzle of platinum. The blowpipe may have a stop-cock to regulate the blast. The lamp has a cotton wick of nearly
an inch in thickness; this is kept together by a tin
wick-holder, which is soldered to the lamp; melted
tallow fills the lamp, and feeds the wick with fuel.
In order to get rid of the smoke, which is in considerable quantity, there is placed on the table, over the
lamp and nozzle, a cover of thin sheet copper. The
fore part of this cover is open, to allow the jet of
flame to pass freely. From the top part two tubes
go upwards for the exit of the smoke; between these,
the operator has a view of the jet of flame and
the object he is at work upon, while his eyes are
screened from the body of the light. The two tubes
join above in one short tube. Over the open end of
this short tube, at a small distance above, is a tube
suspended from the ceiling by wires, which conveys
the smoke into the chimney of the room. By a handle attached to it, the cover with its tubes is removed
whenever it is necessary to trim the wick. The
flame of coal gas may be used instead of a lamp with
a bellows of this kind.—The whole of this apparatus
—the table, bellows, blowpipe, and lamp—is represented by figure 27. The dotted lines at the top of
the figure represent the cover.

46. *Tilley's Hydro-Pneumatic Blowpipe.*—The
inventor of this machine, who was a traveling fancy
glass-blower, was rewarder by the Society of Arts,
with the sum of fifteen Guineas. The following
description is taken from the 31st vol. of the Tran-
sactions of that Society. The inventor states the
whole apparatus, including lamp and case, to weigh
only three pounds and a half, and that it can be furnished complete, even when made of tinned copper,
for £2 12s. 6d. Were it made of tinplate, (white
iron), japanned, it would be cheaper.

47. "The utility of the blowpipe, in the arts, to
raise a great heat on a small object, from the flame

of a lamp, is too well known to require pointing
out. The assaying of minerals, the arts of enamel-
ling, jewellery, soldering metal works, but above all
the blowing of small articles in glass, are purposes
to which it is better adapted, than almost any other
mode of applying heat. The usual manner of producing a stream of air for blowing glass, is, by
means of a small pair of double bellows, fixed beneath a table, and worked by the operator's foot ; a
pipe proceeds from these bellows to the top of the
table, and terminates in a small jet, before which a
lamp is placed, and the flame is blown, by the current of air, upon the object to be heated. The defects of the bellows are, that the stream of air is
not perfectly regular, which causes a wavering of
the flame, so that it does not fall suddenly upon the
object which is to be heated. Mr Tilley's blowpipe
corrects these defects, by using the pressure of a
column of water to regulate the stream of air, and
the supply is furnished from the mouth of the operator, by blowing through a tube.—Fig. 15, is a section of this instrument ; and fig. 16, shows a per-
spective view of it in action. A A, is a vessel of
tinned iron, or copper, about 17 inches high, 5 wide,
and 9 broad ; the lid of which opens on hinges, and
supports the lamp, B, which burns tallow instead of
oil. C is the blowing-pipe, by which the air is
thrown into the vessel ; this, as shown in the section, fig. 15, has an inclined partition, D, which divides it into two chambers, E and F ; but, as the
partition does not reach to the bottom of the vessel,
the two compartments communicate with each other
underneath it : that marked F, is closed at the top,
so as to be air-tight ; but the other is only covered
by the lid of the vessel, and may therefore be considered as being open to the outward air. The pipe
C, is soldered air-tight, where it passes through the
top of the chamber, and descends very nearly to the
bottom of the vessel, deeper than the partition, D,

does; so that its mouth is always immersed beneath the water. The metallic part of the blowpipe, G, which conveys the blast of air to the flame of the lamp, is likewise soldered into the top of the chamber, F; it holds a bent glass tube, a, which terminates in a very small and delicate jet, and is fitted air-tight into the tin or copper tube, G. Now, by blowing into the tube, C, the air is forced out at the bottom of it, and rises in bubbles through the water into the upper part of the chamber, F; this displaces a corresponding quantity of water, which passes under the partition, D, into the other chamber, E, elevating the surface of the one column of water, and depressing the other, as shown in the figure; the water endeavouring to return to its original level, causes a constant compressure of the air and forces it through the jet, a, into the flame of the lamp. By this means, it is not necessary to blow constantly with the mouth; for, though the air is forced into the receiver at intervals, yet the pressure of the water will expel it in a constant stream; and the operator will not be fatigued by the motion of the foot, necessary in working bellows; nor need even to keep his mouth at the pipe constantly, but merely to blow from time to time, as he finds the stream of air to decrease in its power. A, must be three parts filled with water.

48. "The metal socket, G, which connects the glass tube or blowpipe, a, with the vessel A, is made conical; and the tube, having a piece of paper first wrapped round it, is bound round with cotton wick-yarn, in a conical form, so as to fit the socket tight, and yet permit the tube to be moved in any required direction, to cause the air to act properly upon the flame; and the curved metal tube, C, is also fixed into the other part of the same tube, in a similar manner. H H are the two sides of a tin frame, which is fixed in front of the vessels; and has grooves withinside of them, to receive a tin plate, I,

which forms a screen; and can be adjusted in height, so as to keep the light of the lamp from the operator's eye; though he can see the work over the top of it: this screen is held fast, by its foot being placed between the lid of the vessel and the top of the close chamber, F.—K is one of two handles, which support the operator's arms, whilst holding a glass tube, or any other article, in the flame; and there is another like it, at the opposite side of the vessel: these handles are wrapped round with woollen list, or leather, so as to form cushions; and the vessel is steadily fixed upon a chair or bench, by means of a leather strap, buckled to the loops on each side of it, and passing underneath the chair.

49. " The lamp is made of tin; it is of an elliptical, or rather, of a bean, or kidney shape, one side being curved inwards; across the centre of it, stands a metal wick-holder; having a loop on one side of it, which is soldered to the bottom of the lamp; see fig. 17. Through this loop, the wick; of cotton-yarn, is drawn; and being cut, and then opened both ways, as shown in that figure and in fig. 18, forms a passage in its middle, through which the current of air from the jet, a, passes; as in fig. 15, 16, 18; and carries the long pointed flame upon the object to be heated. The lamp, figs. 17, 18, is filled with tallow; which, melting by the heat, becomes fluid, and burns as well as oil, but with a less offensive smell; and, when cold, being solid, is more conveniently carried than oil. This lamp is placed within another vessel or stand, marked B, in fig. 15, 16, 18, which supports it at a proper height; a space being left between them, all round the lamp, to receive any tallow which may run over its edge.

50. " In using this blowpipe, the following observations being attended to, will greatly increase its effect. The long, flat, cotton wick of the lamp, will be found to act better than the usual round cotton-wick; but, in either case, the flame which is raised will be considerable. The end of the

glass pipe or jet, *a*, must be just entered within the flame, and the current of air will throw out a cone of flame from the opposite side. If it is properly managed, this cone will be distinct, and well defined, and extended to a considerable length. Care must be taken, that the stream of air does not strike against any part of the wick, as it would then be disturbed, and the cone split into several parts. (A wire pointed at its end, and bent, as shown at fig. 19, is very convenient to smooth the air passage through the wick) ; the jet of air must be delivered somewhat above the wick ; and as, unless the flame was considerable, there would not be sufficient for the stream of air to act upon, for this reason the wick is opened, as shown in fig. 17, that it may expose the largest surface, and produce the greatest flame; the stream of air from the pipe should be directed through the channel or opening between the divided wick, so as to produce a cone the most perfect and brilliant. On examining this cone of flame, it appears to be of two different colours, the part nearest to the lamp being of a yellowish white, and that beyond, of a blue or purple colour.

51. " The subject to be heated, is held in the flame, at the termination of the yellowish-white part, where it receives the greatest heat, and is not discoloured by the soot, which accompanies the white flame.

52. " Glass tubes are, when applied to the flame, quickly rendered pliable, and may be bent, or drawn out into threads or points, and hermetically sealed ; or, by blowing into the other end of the tube, it may be expanded into a small globe ; or it may be made to form various other small articles, at the pleasure of the operator.

53. " In chemistry, mineralogy, and the arts, the blowpipe is an extremely useful instrument; being capable of throwing so powerful a heat on a small object, as can only be obtained, on a larger quantity of the same substance, in the most powerful furnaces : and with this advantage, that the process is always under the inspection of the operator ; whereas he can only conjecture what passes in the centre of a furnace.

54. " In using the blowpipe for experiment, a piece of charcoal is generally employed to support the subject, and hold it in the flame of the lamp ; the charcoal should be of a close compact grain, and properly burnt ; for, if it is too little carbonised, it will flame like a piece of wood, and obscure the object ; and if it be too much burnt, it is so quickly consumed and burnt to ashes, that the object is in danger of being lost in it. The charcoal greatly increases

C

the heat, by reverberating the flame, and by heating the
object at the opposite side; it being converted into fuel, and
excited by the blast: thus creating an atmosphere of heat
and flame around it, which prevents the heat from being
carried off so fast, or the object being so much cooled,
if it should, for an instant, be moved out of the cone of
flame, from the unsteadiness of the hand, or from accidental currents of air, which would disturb the flame, and
cause such a wavering in the point of the cone as to divert
it, in some measure, from the object.

55. " In order to prevent more tallow than is necessary,
from being consumed, to produce the intended effect; it is
convenient to have several lamps, with wicks of different
thickness; viz. one to hold two flat cottons (such as are
used for the Liverpool lamps) of about an inch and a quarter broad; another to hold four, and a third six; or as
much common cotton wick-yarn as is equal to these wicks
in bulk.

56. " Glass jets should also be provided, of different sized
apertures, to suit the greater or lesser sized wicks and
flames, and deliver streams of air upon them proportion-
ately: and their jets should, for fancy glass-blowing, point
upwards in a small degree.

57. " Hogs'-lard is equal, or, perhaps, even superior to tallow, for the lamp."

58. *Additional Observations and Improvements on the
above Apparatus, by Mr Gill, editor of the Technical Repository.*—" Having sedulously attended Mr Tilley during
his operations with this valuable instrument; and also considerably assisted Dr C. Taylor, the late worthy Secre-
tary of the Society of Arts, &c. in making out the foregoing description of its construction and uses; besides
constantly having one at hand for his own purposes: the
Editor conceives that it will be desirable to afford his readers such additional information respecting it, as his oppor-
tunities of acquiring it, and his own experience in using
the instrument, have furnished him with.

59. " The greatest inconvenience he experienced, in using
the hydro-pneumatic blowpipe, arose from the frequent
destruction of the glass jets, owing to their points being
fused, on being left, only for an instant, in the flame, after
the air had ceased to pass through them. This inconvenience, at length became so serious, that he determined to
contrive a more facile method of forming them, than by
making each jet of a part of a bent tube as in Mr Tilley's
instrument. With this view, he caused a glass tube to be

bent at a proper angle, which served as a receptacle for all the jets; and which were then readily formed by merely softening a portion of another smaller glass tube, by heat, and drawing it out to the required size; then softening another portion of it, at a proper distance apart, drawing it out, and so proceeding, until as many portions of the tube were so treated as were thought necessary. The tube was then cut into portions (with the help of a proper knife, to be described hereafter), so as to form the jets; and thus a piece of glass tube, long enough to have made only one of Tilley's jets, was formed into five or six of these. This description will be more readily understood, by refering to fig. 28, which represents the bent glass tube, and one of the jets within it. Fig. 29, is a tube contracted by drawing out in the centre where it is intended to be cut asunder so as to form two jets. These jets are readily secured in the bent tube, by merely wrapping a little thin silver or tissue paper round them, which makes a joint that is sufficiently air-tight. Fig. 30, is a section of the knife, mentioned above, for cutting glass tubes with: it is about eight inches long, and tapers to its point; having also a wooden haft or handle fitted to it. This knife must be made of cast steel, and be left quite hard: its cutting-edge is formed by rubbing it across, on both sides, with a shoemaker's *gritstone;* and, in this way, can be renewed, when, in course of use, it has become smooth. This knife is greatly preferable to the three-square saw-files, ordinarily used by the barometer makers, for a similar purpose; or even to the smooth Lancashire *crossing files,* which other fancy glass-blowers employ. In case the nose of the jet is too small, it may be made larger, by cutting off more of the end of it; and it may be further adjusted, by rubbing it across, with a *ragstone,* similar to that used by carpenters, &c.

60. " Fig. 32 is a front view, and fig. 31 an edge view, of a pair of brass tongs, used by Mr Tilley, in the manner to be presently described. They are made of stiff-rolled brass, about one-eighth of an inch thick, and seven inches long; and their circular ends are about two inches in diameter. He used them to flatten the stoppers of smelling-bottles, &c. by pressure, whilst hot, as well as the bottles. They were also used for pressing the red hot ends of glass tubes against, in order to thicken them, previously to blowing into bulbs.

61. " Fig. 33 represents another tin screen, used by Mr Tilley, instead of that shown in fig. 15, 16, at H I, to hide the flame of the lamp from the eyes of the operator; and

yet, at the same time, permit him to have a better view of
his work, than could be had over the latter: it slides up and
down, in a tin socket affixed to the end of the machine
nearest the operator.

62. " Fig. 34 is a cylindrical rod of brass, about five inches
long, and three-eighths of an inch in diameter ; and having
one end tapered to a triangular point. This Mr Tilley
used to widen the orifices of glass tubes, bottles, &c. with ;
and it is exceedingly convenient for that purpose.

63. " It may be desirable to mention Mr Tilley's mode of
forming and adjusting the wick of his lamp, which is as
follows :—Fine cotton wick-yarn is to be wrapped round
the four fingers of the left hand, so as to form a flat coil, of
a sufficient size to fill the loop of the wick-holder, but not
so tightly as to hinder the tallow, from rising freely
through it. A piece of twine must then be passed through
the coil, before taking it off the fingers. The lamp being
next filled with melted tallow, lard, &c., the string is to
be passed through the tallow, &c., and through the loop of
the wick-holder ; and the coil of wick-yarn must be also
drawn through the melted tallow, &c., after it, and be
lodged in the loop, carefully spreading it equally therein.
One end of the wick, viz., that next to the operator, must
then be set fire to ; and, as it burns forwards, the coil ot
wick must be cut open, with the trimming scissors, gradually, until it is all divided, and a passage left through the
middle of it, for the blast of air to pass in : it is also to be
evenly trimmed on each side : it may project, through and
above the loop, about half an inch.

64. " This wick will serve a considerable time, without
snuffing ; provided that, on relighting it, when the tallow
is solid, the end next to the nozzle or jet be only kindled,
and a lump of tallow, lard, &c., be held in the flame. upon
a fork, &c., so as to melt it, and then dropped upon the
forepart of the wick, to feed it, until the tallow, in the
body of the lamp, becomes sufficiently melted, to rise and
furnish it with fuel ; and thus prevent the cotton from being
charred and rendered useless. When, however, it requires
to be renewed, the tallow, &c., being melted, as before
mentioned, the wick may be raised, by means of a pair of
forceps or pliers ; and the charred parts be removed,
with the scissors, by enclosing the wick lengthwise between their blades ; and, bringing away what will readily
come off : and, as in this operation, the wick is partially
closed, it must be again opened and smoothed, by means of
the bent wire, fig. 12, before described.

65. " A slip of oiled silk should be tied over the lower end of the blowing-tube, C, fig. 15, 16, in order to prevent the return of a little water up it, which, otherwise, is apt to take place.

66. " In order to adapt this machine to the purposes of the jeweller, &c., for soldering with; it will be necessary to provide an additional glass tube, like that shown in fig. 28, but with its upper end bent downwards, instead of pointing upwards, as in that figure.

67. " A shallow tin-tray, with its edges turned up, and about four inches broader than the top of the machine, and as much longer, being secured upon it, by means of wire loops, affixed to the tray, and to the machine; viz., one at each side, on the end next to the operator, and another at the further end of it; and through which loops, strings are passed and tied fast; will be found exceedingly convenient to hold pieces of glass tube, &c., and prevent them from falling off the top of the instrument.

68. " Other fancy glass-blowers are in the habit of employing *much thicker wicks* than these described, in order to produce a flame sufficiently powerful for their work, and which they unquestionably do; but then it is attended with a very great waste of tallow, &c.; whereas, in this disposition of the wick, nearly the whole of the flame is carried into the current from the blast, and is usefully employed.

69. " The Editor has lately employed this very useful apparatus, to excite the charcoal fire in his small pumice-stone furnace, by means of the very fine and gentle blast produced from one of the glass jets; and the consequence has been, that the water contained in his portable boiler,* is now heated in half the time formerly required: and so gentle is the blast of air, that none of the light ashes of the charcoal are blown about by it; and it is now become an exceedingly useful and cheap substitute for the oil and spirit lamps, in chemical experiments.

70. " His machine is made of *tinned copper*; and he has found that the water keeps constantly fresh in it, and never requires changing for years together. One singular effect, however, has been produced thereby, which is, the formarion of a considerable quantity of *fine oxide of tin*, very fit for polishing speculums, gold, silver, &c. May not this discovery be useful in the arts ?"—*Technical Repository.*

---

* See Technical Repository, v. 4, p. 247.

71. *Dr. Clarke's Gas (or Oxy-hydrogen) Blowpipe.*
—This apparatus is represented as fitted up for use,
by fig. 21.—1, and 2, represent a deal screen, placed
before, 3, a window, in two parts ; 1, being made to
open like a door ; 2, a fixture before 3, the window.
4, a large bladder or silk balloon varnished with a solution of gum caoutchouc, for containing hydrogen
and oxygen gases mixed in the proportion in which
they combine to form water ; that is to say, two parts,
by bulk of hydrogen, and one of oxygen. 5, the
copper reservoir for containing the mixed gases in a
state of compression. 6, The stop-cock of the jet.
7, the flame produced by the mixed gases during
combustion, as propelled from the mouth of the jet.
8, The handle of the piston belonging to the condensing syringe 11 : this piston is on the outside of the
screen, and is intended to be worked by the operator's
assistant during the whole time that an experiment
is taking place; so as to compress a constant supply
of the mixed gases. 9, a deal tray, the bottom of
which may be covered with paper, glass, or metal, as
required for the different experiments. 10, the extremity of a pair of steel forceps in the position of
holding an assay to the jet. It will be understood,
that, during the time of an experiment, the two
parts of the deal screen, 1 and 2, are shut close, to
prevent any serious effects resulting to the operator
or the looker-on, in case the reservoir were to explode.
—Fig. 25, represents the safety-cylinder for the Gas
Blowpipe invented by Professor Cumming. In the
engraving above described, the safety-cylinder is
seen immediately under 5, in the upper part of the
copper reservoir, where the top of it containing wire
gauze appears ; and where one of the extremities of
the jet is screwed into it. In figure 25, the safety-cylinder is seen as when it is taken out of the reservoir.
A, shews the cap containing the wire gauze ; B, the
stop-cock ; C, the mouth of the jet ; C, D, the volume of the flame ; E, the interior of the cylinder,

shewing the height to which it is filled with oil ;
F, a valve, at the bottom of the cylinder, communicating with the gas reservoir in which the mixed
gases undergo compression ; *x, y,* a wire gauze over
the valve F.

72. By gradually turning the stop-cock of the jet
belonging to the gas blowpipe, the violence of the
gaseous flame may be diminished or increased at
pleasure; and, of course, the degree of heat may be
modified : its utmost intensity being afforded when
the stop-cock is quite open.

73. The operation of exhausting the copper reservoir 5 of common air, and filling it with condensed
gas by means of the syringe 11, and piston 8, is so
simple as not to need description.

74. During the whole time the jet is burning,
the oil will be heard to play in the cylinder. If the
current be inflamed, and the instrument abandoned
to itself, the jet will go on burning until the expan-
sive force of the atmosphere within the box is no longer sufficient to propel a stream of gas with the required rapidity through the tube ; at this time, unless
the bore of the tube be very fine, the flame will pass
backwards, and fire the gas contained in the upper
part of the cylinder; this having exploded, the
effects will cease ; the gas in the reservoir remaining as before—the wire gauze and the oil preventing
the flame retrograding so far as the body of the
reservoir. When the use of the instrument is regularly required, a fresh portion of gas is condensed
into it, before the atmosphere is quite out.

75. Attention should be paid to the quantity of oil
poured into the cylinder: it should cover the gauze,
but not to too great a height.—If there be too much
oil, the agitation caused by the passage of the gas
through it, may cause drops of it to be thrown to the
orifice of the jet, and thereby cause a sputtering in
the flame. Before the apparatus is laid by, the cylinder should be emptied of oil and cleaned.

76. *Goldsworthy Gurney's Oxy-Hydrogen Blow-pipe.*—The construction of this apparatus resulted from Mr Gurney's attempts to improve the gas blow-pipe used by Dr Clarke; the danger attending the employing of which, notwithstanding the alterations of Professor Cumming and others, still remained such as to deter all but a few bold experimentalists from the use of it. The various experiments by which Mr Gurney advanced, step by step, to the object he he had in view, are detailed in his Lectures on Che-mistry. We merely present their results. Fig. 22 represent the apparatus complete: *a* and *b*, is the safety apparatus of which fig. 23 exhibits a section, and through which the gas must pass from the gasometer *d*, by the stop-cock *c*. *g* is a transferring bladder screwed to the stop-cock *h*, by which the gasometer is charged by an assistant during its action, and such a quantity of gas supplied as to keep up a flame for any requisite length of time. Between the gasometer and the charging bladder a valve is placed to prevent a return of the gas, *i* is a wood or pasteboard cap, so contrived as to unite lightness with strength; this is attached by four strings, *k*, to wires, which, passing through holes, *l, l*, in the table of the instrument, are fixed to *m*, a moveable press board below. When the requisite pressure or weight is placed on *m*, the cap *i* is drawn down horizontally and equally on the gasometer *d*; upon which, the gas is forced through the water-tube *b*, safety appa-ratus *a*, and out of the jet *c*, at the end of which it is burned. If an explosion were to happen in the gasometer, the cap *i* would be thrown into the air, where, from its extent of surface and great lightness, its progress would be arrested before any mischief ensued. The gasometer bladder, or silk bag, *d*, is tied to a bladder-piece, which screws into a tube contained in the body of the table of the instrument. This tube terminates in the stop-cocks, *c*, and *h*.

77. 23 is a section of the parts *a, b*, of the prece-

ding figure enlarged. *p* is the stop-cock which admits the gas from the gasometer to the water trough *g*, by a tube *m*, which reaches to the bottom of that vessel. *l* is the water with the gas rising through it. *r* is a gauge which indicates the proper height of the water. *n* is a cork, which, if an explosion happens on the surface of the water, is thrown up, or which can be taken out, when water is to be poured into the trough. *i, i*, are chambers in the safety-apparatus, intended, by means of the wire gauze partitions, *k*, to arrest the progress of a retrograde flame. *o* is the jet, of which various sizes should be provided, to be used at the will of the operator.

78. Fig. 24 represents Mr Wilkinson's improved safety-chamber for the above apparatus, which allows the use of a larger jet, and has thereby increased the range and utility of the instrument. The cylinder *f, f*, is of brass, about three quarters of an inch long and the same diameter internally, and is filled in the following order, beginning at the jet end :—Ten layers of wire gauze, having about 4000 apertures to the square inch, *g*; a layer of asbestus, one-eighth of an inch thick, *h*; ten more layers of wire gauze *g*; another layer of asbestus *h*; twenty layers of wire gauze *g*:—the two ends of the cylinder being concave, in order to afford as large an area as possible for the passage of the gas from the entrance pipe *i*, to the jet *j*. This invention was rewarded by the Society of Arts with a large silver medal.

79. *Miscellaneous Notices of Blowpipes.*—Although we do not think it necessary to trouble the reader with descriptions of any other blowpipes than the preceding, which we believe to be the most useful that have been invented, we may nevertheless, by way of closing this part of the subject, say a few words respecting certain other contrivances, which at various times have been brought before the public.—1. Mr Rob. Hare of Philadelphia contrived a hydrostatic blowpipe, capable of using either com-

mon air or gas, and possessing considerable power.
It is described in *The Philosophical Mag. Vol.* 14.—
2. The Abbé Melograni, of Naples, employed two glass
globes; one of which being half filled with water,
was connected with the other, in such a manner as,
by emptying the water into it, to force out a stream
of air through a lateral aperture. See *Pantalogia*,
and *Nicholson's Journal* vol. 9.—3. Dr Berkenhout
suggested the application of the Irish bagpipe to the
purposes of a blowpipe; this, being filled in the
usual manner, and compressed by the arm, could be
made to produce a blast either strong or weak as
occasion required.—4. A blowpipe, said to be "ingenious and useful, but complicated," the invention
of a Mr Haas, is described in *Nicholson's Journal*
No. 10.—5. In the Bibliotheca Britannica it is stated
that Joseph Hume Esq. M. P. is the inventor of a
blowpipe described in the *Philosophical Mag.* v. 44.
—6. *Brewster's Edinburgh Encyclopedia.* (Art. Blowpipe) contains descriptions of several blowpipes
differing from those described here, but not seeming
to possess superior qualities. The most remarkable
one, is a mouth blowpipe with two jets, one intended
to be pointed *to* the flame—the other *from* it. Most
operators find it quite difficult enough to keep up a
supply of air for one jet; yet the author recommends
this invention as tending to render the labour of
blowing more easy. We confess that we are perfectly blind to the merits of this piece of mechanism.
Besides the blowpipes we have mentioned, there
certainly may be others well worthy of notice, but
none such have fallen under our observation.

THE PROPER KIND OF FLAME OR
COMBUSTIBLE.

80. A considerable diversity of opinion exists with

regard to the material which should be used to produce the flame for the blowpipe: wax, oil, and tallow, have each its champion. If a very large flame be used, it does not easily yield to the blast, and a small one produces a weak effect. Bergman directs us to choose a slender candle, either of wax or tallow, with a cotton wick, which he recommends to be bent, after snuffing, in the direction in which the flame is to be impelled. Engeström recommends, as the most cleanly and convenient, that the candle be made of wax, with a wick thicker than ordinary. Children directs us to take a large wax candle. Gahn used candles at first, but afterwards rejected them, and employed a lamp furnished with a large wick and fed with olive oil. Candles do not always furnish sufficient heat, and, besides, are subject to this inconvenience, that the radiant heat from the substance under examination melts the tallow or wax and occasions them to burn away too fast. Berzelius tells us, that lamps are by far the best, and that the best fuel for them is olive oil. A brass cap, fastened over the top of the lamp, by a screw and a collar of leather will effectually prevent the escape of any oil in traveling.—These observations, with the details given in paragraphs 50—70, respecting the management of the lamp belonging to Tilley's blowpipe, will afford sufficient information on this point.

ON THE METHOD OF BLOWING THROUGH THE TUBE.

81. As it is absolutely necessary, says Bergman, that the air should flow through the orifice in a continued stream as long as the experiment requires it, the lungs will be extremely fatigued by this labour, unless a respiration, equable and uninterrupted, can be continued at the same time. To proceed in this opera-

tion without inconvenience, some labour and practice
are necessary. The whole artifice, however, consists
in this, that, while the air is inspired through the
nostrils, that which is contained in the mouth be
forced out through the tube by the compression of the
cheeks.—To some persons this is difficult; but frequent trials will so establish the habit, that a continued stream of air may be supplied for a quarter of
an hour or more, with little other inconvenience than
that attending the lassitude of the lips which compress the tube.

82. To the above directions of Bergman's, it
may not be amiss to add, for the learner's advantage, a proper arrangement of introductory experiments on the art of blowing so as to keep up a continued blast; that being a matter of so great importance. To breathe with advantage to the system
through the nostrils entirely, with the mouth closed,
is the first step to be attained: this seems so extremely easy, as to be almost unnecessarily mentioned; but it is the duty of the teacher of any art to
begin with the simplest movement. This being completely effected, let the learner transfer the air into
his mouth, allowing his cheeks to distend as the air
arrives through the posterior nostrils, and then let
him make two or three moderate inspirations and
expirations by the nostrils, without opening the lips
or suffering the air to escape from the mouth;—this
also is so easy that a very little practice will complete it. Our learner having effected this, let him
introduce between his lips, the mouth-piece of a
blowpipe having a small aperture, and then having
filled his mouth with air, let him allow the same to
be gently expelled through the tube, by the action
of the muscles of the cheeks; at the same time that
he continues breathing without interruption through
the nostrils. This is done by applying the tongue
to the roof of the mouth, so as to interrupt the communication between the anterior part of the mouth

and the passage of the nostrils. When the mouth begins to be empty, it is replenished by the lungs in an instant, while the tongue is withdrawn from the roof of the mouth, and replaced again in the same manner as in pronouncing the monosyllable *tut*. After practising this for a few days, the muscles employed will be accustomed to this new exertion, and unless the flame be urged too impetuously, a continued current may be produced without any extraordinary exhaustion.

83. Even the painful feeling arising to the lips from a long-continued compression of the tube, may be in a great degree obviated by having the mouth-piece of an oval or flattened, instead of a round shape. It is easy to give this form to glass; and, for the better kinds of blowpipes, mouth-pieces of this shape can be procured, made either of ivory or silver.

OF THE BLAST AND FLAME.

84. Having, by the observance of the foregoing directions, accomplished the first object of keeping up a steady blast, the next to be attained is the power of producing a good heat and of regulating its intensity. Suppose you are operating with a candle, which has the wick snuffed rather short, and a little bent at its summit *from* the blast in the manner already directed, the orifice of the blowpipe is to be held just above the bent wick, and the air gently and equably expressed; so as to be blown along the horizontal part of the wick as near as may be without striking it. At the same time, all casual currents and draughts of wind are to be carefully avoided, as rendering the flame unsteady, and very materially impairing its strength. The above conditions duly complied with, the flame being forced to one side by

D

the violence of the blast, exhibits the figure of a
cone, evidently consisting of two parts, an outer and
inner: the internal figure being of a light blue colour and converging to a point at the distance of
about an inch from the nozzle; the external figure
is brownish, vague, and indetermined. The most
intense degree of heat is excited at the apex of
the blue flame; the outer flame possesses a much
lower degree of heat than the inner. Fig. 39
exhibits the appearance of the flame of a candle
when acted on by the blowpipe. If the flame be
ragged and irregular, it is a proof that the orifice of
the blowpipe is not round and smooth, and if the
flame have a cavity through it, the aperture of the
pipe is too large. When the fault of a blowpipe is
known, the remedy is obvious.

· 85. The following observations are from Berzelius : " As different bodies have different modes of
ignition, and as we are easily deceived by the light
which they emit, we only learn to distinguish the
maximum of heat by practice. To attain this maximum, we must neither blow too violently nor too
gently; in the first case, a portion of the flame is
either carried off or extinguished by the impetuosity
of the current; in the second, the flame flags for
want of a sufficient supply of air. When we wish
to try the fusibility of a body, or to reduce metallic
oxides which have a strong affinity for oxygen, a very
high temperature is necessary. But many operations
require a less intense heat. The most important
point in pyrognostic assays, is the power, *easily acquired,* of producing at will the phenomena of oxidation
and reduction. The former is so easy, that one
need merely be told how it is to be done, to be able
to do it; but reduction requires practice, and a certain knowledge of the different modes of conflagration.

· 86. " *To Produce Oxidation*—The subject under
trial is heated before the extreme point of the flame,
where all the combustible particles soon become sa-

turated with oxygen. Too great a degree of heat must be carefully abstained from. Oxidation goes on most actively at an incipient red heat. The opening in the beak of the blowpipe must be larger for this kind of operation than in other cases.

87. "*For Reduction, or Deoxidation*—A fine beak is to be employed, which must not be inserted too far into the flame of the lamp; by this means we obtain a brilliant flame, the result of an imperfect combustion, which is the best adapted for this purpose, although formerly the blue flame was erroneously considered the proper one for the reduction of oxides. The flame must be directed on the assay so as to surround it equally on all sides, and defend it from the contact of the air.

88. "A very advantageous mode of acquiring the art of making a good flame for the purpose of reduction, is to keep a small grain of tin in a state of fusion, at a reddish-white heat, on a piece of charcoal, preserving the metallic brilliancy on its surface. In this experiment, it is easy to learn when the flame ceases to be a reducing, and becomes an oxidating one; for the instant this takes place, the tendency which tin has to oxidation, occasions it to lose its metallic brilliancy and become covered with a crust of oxide. When a small grain of tin can be easily kept in a state of reduction, the size of the assay may be increased. The larger the quantity of tin which he can thus keep in the metallic state at a high temperature, the more expert is the operator in his art."

89. Almost all authors who have written on the blowpipe, when they arrived at the part relating to the blast and the management of the flame, have thought it incumbent upon them to say something relative to *the nature of flame*. Among others, Berzelius tells us, that " to produce a good heat, requires some knowledge of flame, and of its different parts," and then he gives an anatomical description of flame, to which Mr Children has added very long and very profound notes from Davy and others, to afford " correct

notions respecting flame." We found all these explanations very tiresome, and think them very useless. In our opinion, " correct notions respecting flame," such at least as can be got at the present time, are not, to the operator with the blowpipe, worth the trouble of acquiring. If it was really necessary to be acquainted with the nature of flame, to be able to produce a good heat with the blowpipe, we fancy that that instrument would soon return to its original insignificance: the information we have on the subject is so little, and that little so wrapped up in the theories, and disfigured by the contradictory statements, of rival enquirers, as to be very difficult to obtain, and almost useless when obtained. But happily the practical use of the blowpipe does by no means depend upon the operator's *opinions* of the composition of flame; and, therefore, we think it useless to enter into lengthy disquisitions of a nature foreign to our subject: it is sufficient for the purpose, if the student learns *where* the hottest part of the flame is, and what effect various parts of the flame have upon the assay; —*necessity* obliges him to go no farther.

90. The enquiry into the nature of flame, is, however, a curious speculation, and very singular are certain solutions of this enigma which have been offered by different " philosophers." One asserts that flame is opaque, another declares it to be transparent; a third is convinced that the flame is increased by the combustion of the air of the jet, while a fourth is equally certain, that the blast merely acts mechanically by battering the flame together. He who may choose to occupy himself with the examination of these statements, will become farther acquainted with the subject by applying to the writings of Rumford, Davy, Sym, Poxrett, Gurney, and Berzelius.

APPARATUS USED WITH THE BLOWPIPE.

91. We purpose describing under this head, a variety of instruments, which, if not particularly connected with the employment of the blowpipe, are at least useful in the study of mineralogy. As the uses of each will be detailed, the student will be enabled to select from the whole, those which the extent to

which he may wish to carry his researches, may render necessary.

92. The following sentence is from Berzelius: "Order in the arrangement of the various instruments is very essential, so that the operator may in a moment lay his hand upon whatever he may want." To this passage, which contains an indisputable truth, Mr Children has affixed the following note :—" Here follows a long detailed description of a table, with a drawer at each side, and four in front, divided into moveable compartments of tinned iron, to hold the various instruments, &c. not forgetting *a hook with a towel fixed to the right leg of the table!* Next comes an equally elaborate description of a red morocco case to hold a traveling blowpipe apparatus. These things are all very useful, but I cannot agree with my author, that a particular description of them is necessary. I have, therefore, omitted them." Not contented with thus turning into burlesque, what, in the opinion of the experienced Berzelius, was deserving of serious consideration, Mr Children has another farcical touch in his preface :—" To the author," says he, " some apology perhaps is necessary, as to certain liberties I have taken with the original. In the first place, I found the description of apparatus so very minute, that though such may be desirable in Sweden, in England I am sure it is not wanted, abounding as this country does in skilful artists, from whom every species of philosophical apparatus may be had, of the best workmanship and construction. I have, therefore, shortened several of these descriptions, without however omitting any thing essential." We like the *naïveté* of these passages particularly. Coming from an *English Philosopher*, they are equally remarkable for drollery and modesty. Mr Children opines that a very minute description of blowpipe apparatus, though useless in *England*, may be desirable in *Sweden*. How strikingly must every one differ in opinion from Mr C. who knows, that, for almost all the information we possess respecting the blowpipe, we are indebted to Sweden ! The use of the instrument originated in Sweden ; it was improved in Sweden ; it was brought to perfection in Sweden. The most eminent individuals who have used the blowpipe, such as Swab, Cronstedt, Bergman, Gahn, and Berzelius, were all of Sweden. While in England—in England we adopt what in Sweden they invent ! What is indicated by the fact that Mr Children translated Berzelius's Work on the Blowpipe into English ?—What but, that, " abounding as this

country does in skilful artists," it was necessary to send to
Sweden for an instructor for them? Yet Mr Children
thinks, that that information which is useful in Sweden is
unnecessary in England—an idea certainly very extraordinary.

93. In our opinion, Mr Children, by mutilating and
omitting Berzelius's description of apparatus, has very se-
riously diminished the utility of the work; and in this opinion we are joined by many persons who like us, place a
high and just value upon practical details, and wish the
expurgated parts had been retained. The fact is, that Mr
Children seems to have thought that all the readers of his
book were to live next door to " Mr Newman, of Lisle
Street, Leicester Square, who," says he, " makes complete
sets of apparatus for the blowpipe, containing every thing
that the operator can require." If Mr C. has need of
any instrument, he can probably procure it from Mr N.
in a few minutes; but many persons are far differently situated. We ourselves, for instance, reside four hundred miles
from Leicester Square, and are ignorant of the existence of
any " skilful artist," by whom we could be supplied with
" complete sets of instruments for the blowpipe" at a less
distance. It is therefore of primary importance to us, that
we should be supplied with descriptions of apparatus so
minute as not to be misunderstood by the merest artificer,
in order that the want of these " skilful artists" of whom
Mr C. speaks, may not interrupt our pursuits and frustrate
our projects.

94. Had we translated Berzelius's work, his descriptions
of apparatus would all have been retained, even to the muc
despised " hook with a towel fixed to the right leg of the
table." In matters of this kind, we should have relied
more upon Berzelius's judgment than upon our own; and
we should have been cautious of praising England for its
proficiency in this art at the expense of Sweden.

1.—PROPER SUPPORTS.

95. By *the support* is meant the substance or instrument which serves to hold the ASSAY (or object
submitted to experiment) to the flame. This must
necessarily be a solid body; it ought likewise to be
of such a nature, as not to combine chemically with
the substance it supports, and, at the same time,

should be so refractory, as not to give way under the heat. The supports in use at present are of two kinds, combustible and incombustible. The combustible support is charcoal, of which we shall speak at length presently. The incombustible supports are Metal, Glass, and Earth ; in the use of all which one general caution may be given : *to make them as little bulky as possible.* The support always abstracts more or less of the heat; and if care be not taken, it may in many cases entirely prevent the flame producing its proper effect on the assay.

96. *The Best Support is Charcoal.*—It should be a compact kind, of even texture, free from knots, and well burnt. Straight pieces should be selected. Charcoal that is porous, or splits in use, or burns with smoke or flame, is unfit for the purpose. Bergman recommended charcoal made of beach or fir. Gahn thought box-wood would yield the best, but never had an opportunity of trying it. Berzelius denounces fir as liable to crackle and scintillate, and to scatter the assay ; and recommends in its stead, that from a sound, well-grown pine tree, or from the willow. Children says box-wood is too good a conductor of heat, and is apt to split. In his opinion, the best charcoal for the use of the blowpipe, is that which is made from alder, and is employed in England in the manufacture of coarse gunpowder. The charcoal should be divided by a saw into paralellopipedons four or five inches long for use. Some of the advantages possessed by charcoal, as a support, are mentioned in paragraph 54. A small hole is to be made near the end of the charcoal, by means of a slip of plate iron bent longitudinally, to receive the substance which is to be submitted to examination. This will prevent those particles which are small and light from being carried off by the blast of air. Charcoal is chiefly employed for supporting metallic substances, especially fragments of ores intended to be reduced : because it attracts the oxygen from the

metallic oxide, and thereby accelerates its reduction to the metallic form. A metal thus reduced may be kept in fusion on the charcoal, which prevents or retards its again attracting oxygen. It is also employed for roasting metallic sulphurets, and arsenical alloys. Dr Ure, in his abstract of Gahn's experiments, tells us, that the sides, not the ends, of the fibres must be used ; otherwise the assay will spread about, and a round bead will not be formed. But, Berzelius, in the following passage, apparently gives us directions quite contrary : " In order to fix the flux to a point on the surface of the support, one of the ends perpendicular to the layers of the wood is to be chosen for its receptacle, if placed on the section parallel to the layers, it will spread over the surface."—If the piece of charcoal you may be operating with, should shew a disposition to crack, it must be heated gradually till it ceases to do so, before any assay is placed upon it. If this is not attended to, but a strong flame is applied immediately, small pieces will fly in the face and eyes of the operator, and carry along with them the matter intended to be assayed.

97. *Metallic Supports*—In those cases where the reducing effect of a charcoal support would be injurious, or when the subject under examination is absorbed by it, a metallic support is to be used. Bergman recommended a small hemispherical spoon, made of silver or gold. But, since his time, vast improvements have been made in the method of working platinum, a metal, which, on account of its high degree of infusibility, the slowness with which it conducts heat, and its quality of resisting the action of many chemical agents, is much superior as a support to all other metals, and is now generally employed. Substances in the metallic state, or those oxides which are reducible *per se* before the blowpipe, must not be supported on platinum—as the support and assay would be fused into an *alloy*. Charcoal is the proper support in those cases. Pla-

tinum is also unfit to be used with certain fluxes, as will be hereafter mentioned. The following paragraphs describe the different forms in which the metal is made use of.

98. *Platinum Spoon.*—Fig. 44 represents a platinum spoon. The bowl is hemispherical, made very thin, and about a quarter of an inch in diameter. The platinum shank is very short, and is soldered to a longer one of silver, which fits into a handle of wood. When a very intense heat is required, the bowl of the spoon may be adapted to a hole in the charcoal. In operating, it is proper to direct the flame of the blowpipe to that part of the spoon which supports the assay and not to apply it immediately to the substance itself. The spoon is a support which is easily managed by the learner, but it is getting into disrepute with those most versant in the blowpipe art. Because, as the quantity of metal necessary to form a spoon, however thin and small it may be, is such as to absorb a great quantity of the heat of the jet, its form is objectionable and disadvantageous. Spoons were chiefly used when a mineral was to be heated with soda; but, it has been found, says Berzelius, that mineral substances may be heated with that flux, better on charcoal than on any other kind of support; so that spoons are become almost superfluous in experiments with the blowpipe.

99. *Platinum Foil*—Was first employed by Dr Wollaston, and has been adopted and approved of by every other experimentalist. The platinum foil must be about the thickness of common writing paper, and cut into slips about two inches long and half an inch wide. This metal is so bad a conductor of heat, that one of these short slips of foil, may be held in the fingers by one end, while it is heated intensely at the other. Berzelius directs us, when we wish to heat and oxidate at the same time, to direct the flame against the lower surface of the foil. Aikin speaks of the use of this support as fol-

lows : " For fusible earthy minerals, and for the infusible ones when fluxes are used, leaf platinum will
be found the most convenient ; it may be folded like
paper into any desirable form, and the result of the
experiment may be obtained simply by unfolding
the leaf in which it was wrapped up."

100. *Platinum Wire.*—Berzelius tells us, that Gahn,
being unacquainted with platinum foil, and disliking
spoons, made use of a platinum wire, two inches and
a half long, and hooked at one end ; as represented
by fig. 35. This apparatus, says he, so completely
answers the purpose, as, in many cases, to be preferable both to foil and charcoal. The manner of
using it is as follows :—The flux is made to adhere
to the hook by moisture, and is fused to a globule ;
the assay is then moistened, made to adhere to the
flux, and heated with it ; by which means an insu-
lated mass is obtained which can be examined very
conveniently. Oxidations, and those reductions in
which the object sought is change of colour, can
also be performed thus. The globule can easily
be detached by a gentle tap against the wire ; and
may either be cooled suddenly by being thrown on
some cold body, or allowed to cool gradually. The
operator will find it advisable to provide himself
with several of these wires. When he is traveling
where good charcoal can neither be met with, nor
can be conveniently carried, he will find them extremely useful.

101. *Cyanite.*—" Saussure," says Dr Thomson,
" cemented a very minute portion of the mineral to
be examined on the point of a fine splinter of cyanite.
By this contrivance he was enabled to make his experiments upon very minute particles ; and this enabled him to fuse many bodies formerly considered
as infusible."—" Cyanite," says Berzelius, " is in no
respect preferable to platinum foil, and has the dis-
advantage of being acted on by fluxes."

102. *Plates of Mica*—may be used in roasting ores,

wherever it is anticipated, that the reducing effect of the charcoal would be injurious. This is recommended by Berzelius.

103. *Glass.*—M. Dodun, in the *Journal Physique*, for July, 1787, recommends a new support of his invention: it consists of a solid piece of glass, of a triangular form, two or three inches long, gradually tapering from its base, the sides of which measure one third of an inch, to a fine point. He moistens this point and takes up a minute fragment, or small quantity of the powdered matter of the substance to be examined. With this apparatus he fused many substances before deemed infusible *per se.* M. Dodun however deceived himself: the fusions which he considered to be *per se,* resulted from the operation of the glass of the support, which acted as a flux. Arthur Aikin speaks of the use of glass supports as follows: " If the mineral to be examined is of a long or fibrous shape, one end may be cemented to the top of the glass rod by heating it; and in this state it may be examined with great convenience."

104. *Small Plates of Clay*—prepared as directed below, are found to be a very useful addition to the blowpipe apparatus. The colours of bodies melted with borax, &c. are shewn to great advantage on them ; and quantities of matter too minute to be tried on charcoal, or in the platinum spoon, may on them be readily examined, either alone, or with fluxes.—*Process for forming the Clay Plates*—Extend a white refractory clay, by blows with the hammer, between the folds of a piece of paper, in the same manner as gold is extended between skins. Then, the clay and paper together, must be cut with scissars into pieces about half an inch long, and a quarter of an inch wide, and afterwards hardened in the fire in a tobacco pipe.

105. *Glass Tubes.*—Fig. 39, represents a little apparatus sometimes used by Berzelius, to perform the

operation of roasting. It consists of a glass tube, two inches long, one eighth of an inch in diameter, and open at both ends. The assay is placed in this tube at a short distance from the end, and is then exposed to the flame in the manner shewn by the figure. Volatile substances not permanently gaseous, sublime and condense in the upper part of the tube, whence they may easily be taken and examined

106. *Matrass.*—This, which is represented by fig. 36, is used to heat a mineral in, which decrepitates; for ascertaining the presence of water in a mineral; or for subliming sulphur, arsenic, or other volatile bodies. It may be heated by the jet, or held over the flame of a spirit lamp. Vessels of this description are also very useful for performing various solutions by the aid of heat, to which it is sometimes necessary to have recourse. Several different sizes should be kept at hand.

107. *Forceps.*—" For non-metallic minerals that are not very fusible," says Aikin, " the best possible support is a pair of slender forceps of brass pointed with platinum." Fig. 42, 43, represent the form of a pair of forceps described by Berzelius, and by that competent authority declared to be on the best construction, and to answer their purpose completely. *ab, ab,* are two thin plates of steel, joined in the middle to an iron plate, *e, e,* hardened at the ends, *a, a,* that they may not be *battered* when used to detach a particle for fusion from the mineral to be examined, and having a piece of platinum, *c,* riveted on the other extremity of each. The platinum points are made small that they may not carry from the assay too much of the heat of the jet. They are closed together by the spring of the steel blades, to which they are fastened, and are opened by pressing against the buttons, *d, d.*

108. Fig. 41, represents the extremities of a pair of forceps invented by Mr Pepys, and intended

to be used when a substance is to be examined which exhibits the phenomena of decrepitation. This serviceable little instrument differs from common forceps in having upon the points a pair of hemispherical caps like earpicks: any small gem or substance liable to be dispersed by the application of heat, can be inserted into this little cavity and examined very conveniently.

2.—ADDITIONAL INSTRUMENTS.

109. *Hammers.*—Berzelius says, two, of hardened steel, are necessary. One having a polished face for flattening the grains of reduced metal, the other having square and very sharp edges for chipping off small portions from a specimen for examination. Fig. 46, represents a hammer of the form used by Bergman. In Phillips's Mineralogy, third edition, may be seen figures of several sorts of hammers recommended by Dr MacCulloch and others, for various purposes in Mineralogy and Geology. The hammers made at Sheffield are better steeled, and finished in a finer manner, than those manufactured elsewhere.

110. *Anvil.*—Bergman recommended, for the pounding of ores after roasting, a small square steel plate and the hammer fig. 46, as a sufficient apparatus, the particles being prevented from being dispersed by an iron ring, similar to fig. 47, within which the substance to be broken was to be put. But Berzelius says, this iron ring does not answer the purpose, and is unnecessary; because the dispersion of the fragments may be prevented by wrapping the mineral to be crushed in a piece of paper. Instead, also, of the steel plate, he recommends as most convenient for the purpose, a paralellopipedon of steel, about 3 inches long, 1 inch thick, and 5-4ths of an inch wide, polished on all its surfaces. The anvil used by the watch-makers in England, which

E

is a plate of hardened steel about an inch and a half
square, and 1-8th of an inch thick, mounted on a
piece of wood half an inch thick, and enclosed in a tin
case, makes an anvil exceedingly convenient for the
traveling mineralogist.   It is sold by those who supply artists' tools for about eighteen pence.

111. *Pestle and Mortar.*—Berzelius recommends
one of agate. The smaller the better. That which
I use, says he, is scarcely two inches in diameter,
and half an inch high on the outside.   Its cavity is
5-16ths of an inch less than these dimensions.   For
a great many purposes, where the utmost nicety is
not requisite, the small Wedgewood's-ware pestle
and mortar, which is a very cheap article, is highly
convenient. The traces left on the surface of the
mortar by metalliferous substances may be removed
by grinding with pumice stone. Many operators
use a mortar made of steel.

112. *Files*—are very useful.   Berzelius recommends "a triangular, a flat, and a round or half-
round file."   A file made at Sheffield, about six
inches long in the blade, and of the shape represented by the section fig. 30, we have found very handy.
Files are used to try the hardness of minerals, to cut
glass tubes, and for a variety of other purposes.

113. *Knife.*—This is one of the most indispensable articles of the blowpipe apparatus. Like the
file it is employed to try the hardness of minerals,
which is estimated by the greater or less resistance
they oppose to it.   "The most unexceptionable method of ascertaining the hardness of a mineral," says
Aikin, "is the greater or less ease with which it yields
to the point or edge of a knife of hardened steel.
Whereas, the comparative ease and vivacity with
which a mineral gives sparks with steel is a very
equivocal test."   The point of the knife is also used
to take up portions of the fluxes, and to knead them
with the pulverised mineral.

114. *Magnifying Lens.*—In the examination of

minerals by the blowpipe, the microscope is often indispensable in ascertaining the result. Until we have examined the object by its means, says Berzelius, we must be cautious how we decide as to colour; for the light which charcoal reflects on small glassy globules, often produces false appearances which the microscope corrects. Magellan recommends, " for the examination of the structure and metallic parts of ores, a triple magnifier, which, differently combined, produces seven magnifying powers." Berzelius prescribes " a microscope with one or two plano-convex lenses of different powers, which Dr Wollaston has shewn is the form best adapted to enlarge the field of distinct vision." When a microscope of the latter description cannot conveniently be procured, a common pocket magnifying glass, with one or two good convex lenses, will be found very convenient. It is proper to examine a mineral with the glass as well *before* assaying as *after*; to be enabled to separate more easily the different parts of it, should it be of a heterogeneous nature.

115. *A Dropping Tube.*—This is a neat contrivance of Dr Wollaston's, for dropping water. It consists of a glass tube with a fine orifice, passing through a cork fitted in the neck of a bottle half filled with water. The tube may be bent at a right angle, about an inch from the orifice. By holding the bottle in the hand, the application of the warmth dilates the air contained in it, and thereby slowly expels the water. There is another contrivance for dropping water, which is a narrow glass tube about six inches long, with a bulb about an inch in diameter blown near the upper end of it, and having its lower end drawn out to a capillary point. The manner of using this is as follows:—while the lower end is immersed in water, the bulb is filled by the action of the mouth applied to the upper extremity. The latter is then closed by the finger, and the water remains suspended in the tube, until, by cau-

tiously removing the finger, it is expelled in drops. The dropping tube is used to wash filters, and precipitates in general; it is also convenient for washing the charcoal powder from the grains of reduced metal, in some experiments.

116. *Electrometer.*—" Electricity," says Aikin, " is a (mineralogical) character of but small importance. It serves indeed to point out the Tourmaline and other pyro-electric minerals, and is connected with some curious and important crystallographical facts; it must not therefore be entirely neglected." We determine the existence of this curious property in a mineral, by means of a little instrument which consists of an upright glass stem four or five inches long, fixed in a wooden foot, and having suspended from its upper end, half an inch of which is bent at a right angle, a very small piece of gold paper, by means of a thread of silk. This instrument is generally called an electrometer, sometimes, more properly, an electroscope. When we wish to ascertain the electricity of a substance, we first charge the gold paper, either *positively* by touching it with an excited glass tube, or *negatively* by means of an excited stick of sealing wax. The gold paper will then be *attracted* by a substance possessing a different kind of electricity, or *repelled* by a body having the same kind as itself. Minerals which become electric by heat, previously to their being presented to the electroscope, are held to the flame of a candle by means of a forceps having a handle of glass. The Tourmaline when operated upon in this manner is found to possess two distinct electric poles, one positive, the other negative; which poles, upon the minerals becoming cold, are reversed. Fig. 48, represents an electroscope contrived by Haüy. It is better adapted for traveling than the simple instrument above described. *a b* is a needle of brass, terminated at each extremity by a ball of the same metal, and moving very easily on

a pivot at the centre. *c* is the metallic base of the instrument. *d* is an insulating plate of resin or glass. When a mineral is to be tried, it is necessary, in the first case, to communicate a certain kind of electricity to the instrument, which is done as follows:—Having excited a tube of glass, or a stick of sealing wax, place a finger on *c*, the metallic base of the electrometer, and then bring the excited body (represented by *e*) within a small distance of one of the balls of the needle. When the needle is sufficiently electrified, *first* withdraw the finger, and then remove the glass tube or sealing wax. If now, a mineral, excited either by friction or heat, be presented to the needle, attraction or repulsion will ensue; and, the electricity of the needle being known, that of the mineral is determined with ease.

117. *Magnetic Apparatus.*—" Few minerals in their native state," says Aikin, " affect the magnetic needle, but a considerable number do so after being subjected to the action of the blowpipe. This character distinguishes pretty accurately the ores of iron from other substances." Fig. 40, represents the magnetic needle of Haüy. The pivot and stand *c*, fig. 48, serve as a common support to this and the electric needle. When the magnetic needle is suspended on the point so as to turn freely, it takes its position in the magnetic meridian, and is prepared to indicate the presence of iron in any substance presented to it. Magnetic pyrites and some other substances attract either end of the magnet to which they are presented; but some ores of iron possess polarity, that is to say, the power of attracting one and repelling the other of the poles.

118. The quantity of iron contained in some minerals is so small, as only to be rendered perceptible by a particular mode of management. This consists in placing the needle in such a manner as to cause it to obey very readily the weakest attraction. We do this, by making the north pole of a needle,

E 3

A, fig. 49, point either to the east or west, by means
of the attraction of a stronger magnet, B; the latter
being so placed as to possess a power just sufficient
to keep the needle in the position required. If then,
any substance containing iron, the mineral actinolite
for example, be presented to the needle, the test will
be found extremely delicate. Berzelius warns us to
take care in this experiment, not to communicate, in
handling it, any electricity to the mineral examined,
since it would act on the needle as a magnetic power.

119. *Addition to the Table Lamp Furnace.*—Fig.
38 represents a small vice, intended to slide up and
down the pillar of a lamp furnace or retort stand.
m is the opening for the pillar. *k* the screw by
which it is fixed. The mouth *i*, receives the triangle, fig. 37, and is tightened by the screw, *l*. This
triangle is jointed at *a* and *d*, in order that it may be
folded for the convenience of travelers. It is made
of strong iron plate, and the length of its sides is
2½ inches. The sides *bc* and *ba*, contain each four
small holes, intended to receive the ends of wires
bent at right angles; so that the large triangle can
be divided into four smaller ones. This apparatus
is intended to support over the flame of a lamp such
articles as small porcelain capsules, watch glasses,
crucibles, and matrasses; with minerals and solutions required to be heated.

120. *Miscellaneous Articles* recommended by Berzelius and others.—A small pair of scissors. A pair
of small tongs to hold crucibles, &c. when heated
by the spirit lamp. A touchstone, and some assay
needles, made of alloys of gold of different stand-
ards, for trying the purity of gold. A polished
blood stone, on which metallic substances which have
been reduced but not fused, may be rubbed, to ascertain if they possess metallic brilliancy. Strong
cutting pliers for detaching small portions for assay
from mineral specimens. A pa of common brass
forceps, such as fig. 50, for handling minute por-

tions of minerals to be examined, and to take out the small buttons or globules of reduced metal from the fluxes when hot. Fig. 45, represents a small instrument, made of thick brass wire, bent and flattened at the ends, very useful for holding a watch glass, with any solution or substance to be heated, over the flame of a lamp. An assortment of small glass tubes, for making test tubes, matrasses, blowpipes, dropping tubes, and similar apparatus. A spirit lamp. Magellan recommends a piece of black flint as a touchstone, which being rubbed with any metal, if it be gold, the marks will not be corroded by nitric acid.

121. *A Tray*—to catch the particles that may fall from the support during an experiment, and prevent our losing an assay with which we have been a good while occupied, is indispensable. We have already alluded to the use of this in paragraphs 67 and 71. The tray may be made of iron, not tinned, with a border about an inch high, and having its bottom covered with a piece of thick white paper. Those who travel will find a white earthenware dish a good substitute for a tray. To avoid confounding the assay with other matters, it must be emptied after every experiment.

OF THE PREPARATION OF AN ASSAY FOR EXAMINATION BY THE BLOWPIPE.

122. By the term ASSAY, we mean simply the small piece of mineral, or other substance, submitted to the blowpipe for examination. This use of the word originated with Berzelius: its brevity renders it convenient. We have thought it proper to give this explanation here, because we may probably employ, in the course of this work, the word *assay* as a *verb*, and, after the old custom, speak of "assaying" a

mineral. We do not apprehend, however, that any
obscurity will hence arise, if the distinction is borne
in mind.

123. *Preliminary Operations—External Characters.*
—When a mineral substance is to be tried, we are
not to begin immediately with the blowpipe; some
preliminary experiments are first to be made by
which those in the fire may afterwards be directed.
For instance, a stone is not always homogeneous,
or of the same substance throughout, although it
may appear to the eye to be so. A magnifying glass
is therefore necessary to discover the heterogeneous
particles, if there be any; and these ought to be
separated, and every part tried by itself, that the
effects of two different things, examined together,
may not be attributed to one alone. This might
happen with some of the finer mica, which are found
mixed with particles of quartz so small as scarcely
to be perceived by the eye. Trap, calcareous spar,
and felspar, are also found mixed together.

124. It is also proper to try the substance by the
electrometer, and to ascertain if it be acted upon by
the *magnet*: because the Electricity and Magnetism
of a mineral are sometimes sufficient to mark it so
distinctly as to render chemical operations unnecessary. A reference to our preceding page will explain
how both of these properties are to be tried.

125. Another thing which should also be well attended to, is the *specific gravity* of a mineral, which
forms one of its most distinguishing characters. We
have not thought it proper however to give any directions for ascertaining the specific gravity of minerals;
because our business is chiefly with the internal or
chemical, and not with the external or oryctognostic
characters. Still it is right that we should remind
the student of the propriety of examining a mineral
of whose nature he is ignorant, in the first place, by
means of the external characters; since, in many
cases, these alone are sufficient to determine with

accuracy the nature of the substance which is the
subject of enquiry. But when the indications of
the external characters do not give the information
which is sought for, then it is necessary to have
recourse to the action of chemical tests; and here
it is that the blowpipe, assisted by the action of
fluxes, in the manner which will be presently described, becomes so very useful.   By removing
doubts, destroying uncertainties, and developing a
variety of characteristic phenomena, it so far ex-
plains the general nature of a mineral, that the discovery of its name and place in a system, becomes
extremely easy.

126. But, supposing the student to have selected
a piece of a mineral apparently homogeneous, and
that the external characters, such as form, colour,
hardness, specific gravity, &c. are insufficient to
point out its name, a circumstance which will very
frequently occur to a young mineralogist, the next
thing to be attended to, if it is intended to try the
mineral by the blowpipe, instead of submitting it to
analysis in the humid way, is

127. *The Proper Size of the Assay.*—If the learner
who wishes for instruction on this point, looks to
the latest work published on the Blowpipe, (that
by Mr Mawe,) he will learn that "the piece of mineral to be examined should not in general be larger
than a peppercorn."   Dr. Ure's Chemical Dictionary gives the same information; and a host of
other writers on Chemistry and Mineralogy, also
describe the requisite quantity to be "the size of
a peppercorn." Bergman, however, with whom
the specification of this bulk originated, observes
that occasionally *smaller* portions must be operated
upon; while, on the other hand, Von Engeström
says, "a piece of about an eighth part of an inch
square is reckoned of a moderate size and fittest
for experiment." There must have been some magic in the term *peppercorn*, to induce writers to make

such use of it; otherwise they would never have employed it so often to denote the size of an assay which could not be generally operated upon. It may be pretty safely asserted that no correct or extended set of experiments on minerals with the blowpipe, was ever made on pieces *nearly so large*; and that no person, using the blowpipe for the first time, on an assay of that size, unless he fell on a very fusible substance, would be able to make any impression on it. It is probable that the failure of many persons who have attempted in vain to use the blowpipe, has resulted from their following the rule of assaying pieces "in general *not larger* than a peppercorn." Nothing can be more evident, than that, if the piece be large, a part of it is necessarily out of the focus of the flame, which is but a small point; and must tend to cool, not only the support, but the part of the assay immersed in the blue apex of the flame. The consequence of which is, that the heat, as it is excited, is carried off; and the operator is exhausted before the assay is in the least affected. The following observations by Aikin, and Berzelius, give more proper ideas on this point than are commonly met with.—" With regard to the magnitude of the specimens (of minerals) required for examination," says Aikin, " no very precise rule can be given: the most fusible, such as some of the metallic ores, may be as large as a small pea, while the more refractory of the earthy minerals should scarcely exceed the bulk of a pin's head." " The morsel," says Berzelius, " is large enough, if we can distinctly see the effects produced on it. And *we are more likely to fail in our object by using too large, rather than too small a piece*. The pieces recommended by Von Engeström and Bergman are many times too large. A piece of the size of a large grain of mustard seed is almost always sufficient. Even if we could succeed in applying an adequate heat to so considerable a bulk as a peppercorn, after all, we can judge of the colour and fusibility, just as well with a smaller piece, which requires neither so strong nor so long a blast."

128. *Of the Figure of the Assay.*—The figure of the small piece which we break from a stone for examination is of little consequence; but it ought to be broken as thin as possible, at least at the edges; for then, the fire having more influence upon the subject, the experiment will be more quickly made. This is particularly necessary to be observed where those minerals are to be assayed which are difficultly fusible; because they may thus be readily brought into fusion, at least at their edges, whereas, if the piece had been thick, this would have been difficult.—*Engeström.*

129. When the assay can only be obtained in small grains, Berzelius recommends them to be placed on charcoal, notwithstanding their liability to be dispersed by the jet. But when it is possible, it is best to select, or to strike from the specimen by the hammer, a very thin scale, or pointed or lamellar morsel, which can be held by the forceps or on the glass rod. And as such a fragment has generally a point or edge very thin or transparent, it is easy to learn, by exposing this to the flame, whether the assay be fusible or not. Accordingly as this edge or point retains its sharpness, becomes rounded, or melts, the assay is relatively fusible or infusible. It is proper to examine the edge after exposure to the flame by the microscope.

130. Certain substances which are very pulverulent, or difficultly fusible, are best examined by being made into a paste, of which a small portion is baked on the charcoal by the jet into a cake and then examined as directed in the preceding paragraph. Dr. Clarke used olive oil to make his minerals into a paste with, but Berzelius recommends simply a drop of water.

consequence. The flame is then to be blown very gradually towards the assay ; in the beginning, not directly upon, but somewhat above it ; and then it is to be approached, nearer and nearer, until the substance becomes red hot ; when all the danger is over. This will do for the most part ; but there are nevertheless some minerals which, notwithstanding all these precautions, it is almost impossible to keep on the charcoal. Thus, says Engeström, from whom this account of decrepitation is chiefly taken, the Fluors are generally the most difficult ; while, as one of their principal characters is discovered by their effects in the fire *per se*, they necessarily ought to be tried that way. On which account, it is best to bind two flat pieces of charcoal together, cutting a channel along the under side of the uppermost piece, and making a cavity in the bottom piece, under the middle of the channel, to contain the matter to be examined. The flame may then be darted into the channel between the two pieces of charcoal, and the substance, which may be considered to be in a closed furnace, will thus be violently heated.— As the fluor, however, will split and fly about notwithstanding this contrivance, a larger piece must be taken than that directed under the head *size of the assay*, in order to have at least something of it left. Very small, brittle, and powdery substances, which are apt to be carried away from the charcoal by the current of flame, may be secured in the manner above-described, as though they were liable to decrepitate.

134. But if the experiment is to be made upon a stone whose effects one does not want to see in the fire *per se*, but rather to ascertain its habitudes with fluxes, then a piece of it ought to be forced down into melted borax, wherein some part of it will always remain, notwithstanding that its cracking will cause the greater part of it to fly away.

135. Bergman treated decrepitating substances

in the way described above by Engerström: he also
used a glass tube and a spoon with a cover.   Berzelius recommends the matrass represented by fig. 36.
Pepys invented for this purpose the forceps, fig. 41.
Wollaston wraps the substance in platinum foil.
Gurney holds decrepitating bodies to the gas blowpipe in a tube of platinum.

136. *Of the Operation called Flaming.*—Alkaline
earths, yttria, glucina, zircon, the oxides of cerium,
columbium, and titanium, and some other bodies,
form clear glasses with borax, which, although preserving their transparency while cooling, become,
upon being slightly heated by the exterior flame,
opaque and milk-white—or, if heated by an inter-
mitting flame, acquire a colour.   In these cases,
however, the flux must be saturated with the assay.
Bodies containing silica, alumina, the oxides of iron,
manganese, &c. are not affected thus by *flaming*,
except the proportion of those substances is exceedingly small.—*Berzelius.*

GENERAL METHOD OF OPERATING
WITH THE BLOWPIPE.

137. Accordingly as the assay is exposed to the
outer or the *inner* flame, on *charcoal* or on *metal*, *with*
fluxes or *without*, it presents numerous and varied
phenomena,—the whole of which being carefully
noted, form, collectively, the RESULT of the trial to
which the substance has been submitted.   Wherefore, it is indispensable that not the minutest circumstance be left unattended to; for, if a single
phenomenon be left unnoticed, the *result* will be
deficient and inaccurate; and its description may be
such as to suit, not the substance examined, but
some totally different one.   Moreover, unless this
accuracy of observation, and precision of description,

be well attended to, the detection of elements whose
presence was not expected will be extremely rare;
and, although this art is capable of being greatly
improved and extended, yet the careless operator can
neither be expected to promote its general utility,
nor derive from it any personal advantage. We
trust, therefore, that the young analyst will bear
constantly in mind the necessity of observing, with
the utmost exactness, whatever occurs during a process with the blowpipe.

138. When an assay has been properly prepared for
experiment, according to the directions given in the
preceding pages, it is, in the first place, to be gently
heated, alone, in the matrass (106), to ascertain if it produce certain phenomena, which are capable of being
developed at a low temperature. We have thrown the
descriptions of these phenomena into a sort of table,
which we think will be found useful by the student
when he takes up the blowpipe to operate.

———

139. A MINERAL, in the MATRASS, ALONE,

May, or may not,

A—*Give off Water.*

> REMARKS.—1. Sometimes, during decrepi-
> tation, a substance gives off but little wa-
> ter; but, when it has nearly attained a red
> heat, (the point at which decrepitation
> ceases,) affords a pretty large quantity.—2.
> Most *hydrates* retain their water with con-
> siderable force.—3. It is necessary to ob-
> serve the effect which the water, disengaged
> by a mineral, has in changing the colour of
> *test papers.*

B—*Change Colour,*

  i. Wholly,
  ii. Partially.

REMARK.—Some minerals become spotted. This is owing to their containing particles which are unalterable by the heat.

C—*Decrepitate,*

  i. Slightly,
  ii. Violently.

D—*Give off Volatile Matter,*

    Which may be

  i. *Arsenic,* known by an odour resembling that of garlic.
  ii. *Sulphur,* known by the odour of sulphurous acid gas, produced by burning brimstone.
  iii. *Selenium,* known by a very strong and disagreeable odour, resembling that of decayed horse-radish.
  iv. *Mercury,* whose peculiarity of appearance is too well known to need description.

    REMARKS.—1. Particular attention is to be paid to the Colour, Form, and Consistence of the substance which remains after the operation.—2. The matrass, which is used to separate volatile combustibles from the assay by sublimation, must have a small body, or be a glass tube closed at one end; since, were it large enough to allow the air to circulate in it, the disengaged volatile substances might be inflamed.

140. Having noted down such of the above phenomena as the substance examined has developed, it is next to be tried, *alone, upon charcoal, before the blowpipe flame,* to ascertain what its habitudes may be in that respect. The experiment is always to begin by directing upon the assay, *first,* the exterior flame, and, when the effects which that produces have been observed, *then,* the interior blue flame. " Particular care must be taken to observe whether

the matter decrepitates, splits, swells, liquefies, boils, vegetates, changes colour, smokes, is inflamed, becomes oily, or magnetic, &c."—*Bergman.* These phenomena, and their various gradations, form so many useful characters employed by the analytical mineralogist in discovering and distinguishing the various species of mineral compounds.— The number of these phenomena is greatly increased, and the manner of their developement is infinitely varied, by the use of substances termed fluxes; but, as we have not yet treated of the nature of those bodies, we must, in this place, confine ourselves principally to the observation of such characters as are yielded by minerals tried in the heat of the flame *alone*, or, as it is technically termed, *per se.* The principle which induced us to give a tabular view of the effects produced by exposing a mineral to heat in a matrass, actuates us in giving here a similar table of the phenomena developed by a substance in circumstances quite different.

---

141. A MINERAL, ON CHARCOAL, ALONE, EXPOSED TO THE BLOWPIPE FLAME,

May, or may not,

A—*Give off Volatile Matter,*
 After having been gently heated.

> REMARK.—The *presence* of volatile matter is ascertained by holding the assay to the nose immediately upon taking it from the flame: its *nature* is indicated by the characters described at par. 139, D.

B—*Develope the Odour of Sulphur,*
C—*Develope the Odour of Selenium,*
 After roasting by exposure to the oxidating flame.

D—*Develope the Odour of Arsenic,*

After roasting by exposure to the reducing flame.

> REMARK.—The oxidating flame, says Berzelius, renders the odour of sulphur and selenium more sensible, the reducing flame that of arsenic : whence the distinction here made.

E—*Decrepitate,*

i. Slightly,
ii. Violently.

F—*Become Magnetic after Roasting,*

G—*Fuse,*

i. Wholly,
ii. Partially,
iii. Quickly,
iv. Slowly,
v. Assuming a thick pasty appearance,
vi. Assuming the appearance of a liquid.

H—*Intumesce, or Bubble-up,*

i. Slightly,
ii. Violently.

I—*Effervesce, or Sputter,*

i. Slightly,
ii. Violently.

> REMARK.—Effervescence, according to Dr Ure, is occasioned by the liberation of some species of gas; to which cause Berzelius attributes Intumescence also. He owns, however, that these phenomena require investigation. But, in the meantime, they afford characters by which to distinguish substances that in other respects are alike.

K—*Volatilize, or throw off Fumes,*
  i. Slight,
  ii. Copious,
  iii. White,
  iv. Coloured,
      Which either
    a Leave a remainder,
    b Leave no remainder,
    c Condense in a pulverulent form on the support,
    d Do not condense in a pulverulent form on the support.

    REMARK.—The production of the metallic oxides in fine powder, upon the support, during the experiments, is owing to the combustion of the metals at the instant of their revival; as, also, is the tinge of the coloured flames by which the same are accompanied.

L—*Colour the Flame of the Jet.*

M—*Burn,*
  i. With a Flame,
  ii. Without a Flame.

    REMARK.—When *with* a flame, the colour is to be noticed.

N —*Change its Colour,*
  i. Once,
  ii. More than Once.

    REMARKS.—1. Very remarkable changes of colour are produced when fluxes are used. A curious example will be given, when we come to speak of the fluxes, in which oxide of manganese will be exhibited as changing its colour repeatedly during the course of one experiment.—2. The colour acquired, and its intensity, are to be carefully observed; also, at what period of the process the change takes place.—3. As the colour produced is one of the most certain indications of the nature and proportions of the metallic matter contained in the assay, this point must be well attended to.

O—*Be Absorbed by the Support.*

REMARK.—This sometimes takes place after fusion; but, in most cases, by a continuance of the flame, the substance is pumped up, and made to appear again on the surface of the charcoal.

P—*Fuse, but will yield a Result,*

Which may be
  i A Bead of Metal,
  ii. Ashes or Powder,
  iii. A Glassy Globule—
    a Transparent, wholly, partially, or not at all,
    b Filled more or less with air bubbles,
    c Perfectly colourless,
    d Tinged to a greater or less degree with some colour,
    e Homogeneous,
    f Heterogeneous—
  iv. A White Enamel,
    a Smooth,
    b Having the appearance of a frit,
    c Homogeneous,
    d Heterogeneous.

REMARKS.—1. It is to be noticed, whether a glass which appears dull be really opaque, or is rendered apparently so by the air-bubbles it contains.—2. It is also to be observed, whether the appearance which a glassy globule has *when hot*, continues *during* and *after* cooling; or, if any and what

change takes place.—3. Another thing to
be observed is, whether a substance which
has once undergone fusion, continues fusible, or has become incapable of it.

General Observations.—In all the preceding cases, it is necessary to notice, whether
the phenomena are produced by the *interior*
or the *exterior* flame; and whether upon
the immediate application of the flame, or
after having for some time resisted its action.   It is also necessary to notice *the order* in which the different phenomena are
developed; which will rarely be the same
as that in which they are here described.
Often, indeed, several characteristics are
exhibited at the same time, and it requires
a quick eye, and close attention, to observe
and distinguish them.

OF FLUXES AND RE-AGENTS.

142. When we have observed the alteration which
the assay is made to undergo by the mere action of heat,
it is then necessary to examine what farther change
takes place, when it is subjected to other trials with
the addition of FLUXES,—which are substances of
such importance in this art, that it is necessary to
speak of them somewhat in detail.

1.—THE NATURE AND USES OF FLUXES.

143. The term *flux* is applied in chemistry to those
substances, which, when added to metallic ores, and
other mineral bodies, assist their fusion upon exposure to the action of fire.   Thus, either potass or
soda, pure or carbonated, is capable of forming a
fusible compound with all siliceous minerals; as is

borax with such as are aluminous. Fluxes, in short, act upon refractory substances in the *dry* way, in the same manner that fluids act upon solids in the *humid* way; but their action is not uniformly the same—in some cases it is mechanical, in others chemical. Sometimes the acid which is attached to a metallic oxide, and prevents its reduction to the metallic state, is separated by the superior affinity of the alkaline or earthy base of the salt which is used as a flux. In other cases, metallic oxides, which strongly retain their oxygen, are reduced to the metallic state by being fused with inflammable bodies which combine with the oxygen, and sometimes carry it off in the state of gas.

144. In both these cases, the flux is to be considered as acting chemically. But as, in the reduction of a metal, it might frequently be divided into minute particles, which it would be very difficult to collect, it is advisable to have some substance present, capable of being easily melted, and of allowing the metallic particles to subside through it, so that they may conglomerate and form a single button, instead of presenting themselves in scattered grains. Here, then, it is requisite to have a flux which shall act mechanically.

145. The habitudes of minerals with the fluxes, afford much useful information respecting their nature. Thus, siliceous stones never melt alone, but, with borax and soda, form a glass: gold imparts to borax, phosphate of soda, and boracic acid, a ruby colour: silver tinges those fluxes orange colour: copper produces, with the same fluxes, a bluish green pearl: iron tinges them green: tin produces a whitish opaque enamel: antimony affords a hyacinth-coloured glass, partly flies off in white fumes, and deposits a white powder on the support: arsenic diffuses white fumes when heated on charcoal, and produces an odour like that of garlic: cobalt stains a large quantity of borax intensely blue, and forms

with it a blue glass: oxide of manganese has the singular property, under certain circumstances, of repeatedly changing its colour, as may be shewn in the
following manner:—

146. Melt a small quantity of phosphate of
soda, or glass of borax, with the blowpipe flame,
upon a piece of charcoal, and add to it a very
small portion of black oxide of manganese, melt the
mixture together by the inner blue flame, and the
globule will assume a violet or purple colour. Then
fuse it again, and keep it in a melted state for a
longer time, the result of which will be, that the
violet colour will vanish. This being effected,
melt the colourless globule by the exterior flame of
the blowpipe, and the purple colour will re-appear;
but becomes, as before, again destroyed by a longer
continuance of the heat. The smallest particle of
nitre, laid upon the globule, also immediately re-
stores the red colour. If the globule, when colourless, be melted in a silver spoon, or on an iron plate,
or any metallic support, instead of being on the charcoal, the violet colour returns, and will not be again
removed by any continuance of the heat, so long as
it remains on the metallic support.—ACCUM, *Chemical Tests.*

147. Some minerals, such as dioptase, if treated with
fluxes, without having been previously heated red
hot, dissolve with effervescence, which is occasioned
by the disengagement of carbonic acid or water;
whereas, were they heated before being added to the
flux, no such effect would ensue. This is mentioned
for the purpose of shewing the necessity of paying
attention to all the steps of a process, and suffering
nothing to pass unnoticed.

2.—OF TESTS, OR RE-AGENTS.

148. "When, in an analytical pursuit, the object
of inquiry is—What are the elementary parts of a

certain compound?—we expose that compound, under particular circumstances, to the action of certain bodies or powers, which it is expected will chemically act upon it ; and which, when they do, produce changes so obvious to the senses, as to enable us to decide whether the compound does, or does not, contain the principles which it was suspected to contain. [Thus, in experiments performed in the humid way, if we suspect a liquid to contain *iron*, we add to it a few drops of an infusion of galls ; upon which, if iron is present, a dark purple precipitate will be formed. In similar circumstances, if a solution of prussiate of potass is used instead of the infusion of galls, a precipitate of a brilliant blue colour will be the result. And it will be seen in our experiments, that nitrate of cobalt has a remarkable effect upon substances which contain alumina or magnesia.] The bodies or powers, which produce these changes are called *tests* or *re-agents ;* the proper application of which constitutes the chief part of the proceeding called chemical analysis."—*Chemical Recreations.*

149. With the exception of heat, electricity, magnetism, and the action of the fluxes, very little use is made of chemical tests by the operator with the blowpipe, excepting, which is a very advisable mode of procedure, he combines experiments in the humid way with the processes which are peculiar to his own art. In many cases, the use of the blowpipe, on account of the quickness of its action, and the certainty of the indications it produces, is superior to any other kind of analytical operation ; but, in other cases, it is useless, and recourse to an experiment in the wet way is indispensable. In former times, the operators by *fire* and by *water* were fierce rivals, and mutual despisers of each other's plans ; but, philosophers of the present day have perceived, that each mode of proceeding has its advantages, and accordingly they neglect neither, but occasionally make use of both.

G

3. INDIVIDUAL CHARACTERS OF THE FLUXES AND RE-AGENTS.

150. Having given a concise account of the general properties of these agents, we proceed now to state their individual peculiarities, and to describe the method of obtaining them in that high state of purity, in which it is necessary to make use of them, to ensure accuracy in our experiments.

151. The most useful fluxes for the blowpipe are the following:—

1. BORATE OF SODA, commonly called *borax*.
2. SUB-CARBONATE OF SODA,
3. MICROCOSMIC SALT; a compound of phosphoric acid, soda, and ammonia; termed by Berzelius, *Salt of Phosphorus*.

These three substances were employed by Cronstedt, and have been used by every operator from his time to the present. And it is not a little singular, as Berzelius has remarked, that, in the infancy of the art, the best aids to it should at once have been hit upon; yet, so it is, and amid the multitude of substances which have been tried since the above were first made use of, there is not one that is preferable, or even equal, to either of these in its peculiar province. For particular purposes, several other fluxes have indeed been invented; but though it is necessary to have these at hand for occasional use, they are of but little importance, and are not at all to be compared, in respect of utility, with those named above.

152. For the sake of brevity, when we come to describe experiments, these three fluxes will be called simply *borax, soda,* and *mic. salt.*

153. These substances are to be kept ready pulverised, in small phials, and when used, a sufficient quantity for an experiment may be taken up by the point of a small knife, previously moistened by the

mouth to make the powder adhere to it. It is then to
be applied to the palm of the left hand, and kneaded,
so as to form it into a coherent mass, with the addition of a drop of water if required. If the *assay* is
in a pulverulent form, it may be kneaded in the hand
with the flux; but if it be in the form of a spangle or
grain, the flux is first placed on the charcoal, and by
means of the flame, gradually applied, is heated till
it melts into a clear bead. The substance which is
to be examined is immediately to be placed in the
melted flux, and is then ready to be submitted to
the action of the blowpipe.

BORAX.

154. This excellent flux, is of the greatest service
in analyses by the blowpipe. Its use is founded
upon the tendency of its component parts to enter
into composition with mineral bodies by fusion, and
form various kinds of double salts. "Each of its
principles," says Bergman, "is separately fusible,
and each dissolves a great number of other matters."
The whole of the compounds it forms are more or
less fusible, and form glassy globules, the nature of
the colour, and the degree of the transparency of
which, afford indications whereby we can judge of
the nature of the substances which have been submitted to experiment.

155. *To obtain pure Borax.*—Gahn found that in-
fusing borax of commerce with soda on charcoal,
till the two salts were absorbed, there was frequently
obtained a white metal, which was probably derived
from the vessels in which the borax had been prepared from tincal. Borax is also often adulterated with
alum and with fused muriate of soda. To get rid
of the "white metal," Berzelius informs us, it is
merely necessary that the borax should be dissolved
and crystallized afresh. The presence of the other
two foreign bodies may be detected thus :—Dissolve

the borax in water, and saturate its excess of alkali
with nitric acid.   To a portion of this saturated solution, add nitrate of barytes : if alum be present, a precipitate of sulphate of barytes will be formed.   To
another portion, add nitrate of silver : if muriate of
soda be present, muriate of silver will be precipitated.
Chaptal found the simplest method of purifying the
borax of commerce to consist in boiling it strongly,
and for a long time, with water; this forms a solution which by evaporation affords pure crystals.

156. *The way to keep Borax for Experiments with
the Blowpipe,*—is either in the state of small crystals
of the size required for an assay, or in powder, in
which case, as we have before mentioned, it is taken
up by the moistened point of the knife.   "Crystallised borax," says Bergman, " exposed to the flame
upon charcoal, at first becomes opaque, white, and
wonderfully intumescent; it throws out branches,
and various protuberances ; but, when the water is
expelled, it is easily collected into a mass, which,
when well fused, yields a colourless spherule, that
retains its transparency even after cooling."—" Some
operators," we may add from Berzelius, " fuse it, to
get rid of its water of crystallization, and thus avoid
the intumescence which precedes the fusion of the
crystals on the charcoal.   This were a very good
precaution, but that the borax soon recovers its water of crystallization, and swells up before the blowpipe as at first, though in a rather less degree.   *I always use my borax uncalcined*; for the intumescence
is a slight inconvenience, and it is easy to melt the
mass into a globule."

157. The number of substances, the fusion or solution of which we effect by means of borax, is very
considerable, as will at once be seen on a reference
to the experiments in the following pages.   The
chief distinctive characters obtained by treating a
mineral with this flux, are the following: the degree
of its solubility, and whether it be effected quickly

or slowly, with effervescence or without; the colour communicated to the glass; the alteration which takes place with regard to transparency; and whether the globule becomes opaque by flaming (see 136). Of the minutiæ which are to be observed when a mineral is fused with this or any other flux we shall presently give a tabular view; so that we need not enter into particulars here.—With respect to the form of the mineral to be examined with borax, it is better, Berzelius tell us, to begin by attempting the solution of a *spangle* rather than of *powder*.

SODA.

158. In the employment of soda as a flux before the blowpipe, we have two principal objects in view. —The first is, *to ascertain whether bodies combined with it are fusible or infusible :* the second, *to effect the reduction of metallic oxides with greater readiness.* —"In both of these respects," says Berzelius, "it is one of the most indispensable re-agents." Speaking, also, in another place, of the reduction of metallic oxides by means of soda, Berzelius remarks, "This mode of assaying, by which we often discover portions of reducible metal so minute as to escape detection by the best analyses made in the moist way, is, in my opinion, the most important discovery that Gahn ever made in the science of the blowpipe."

159. *To obtain Soda.*—There are two distinct compounds of carbonic acid and soda; the one containing precisely half as much carbonic acid as the other. We may use indifferently either the one or the other for the blowpipe; but the first, or sub-carbonate is generally preferred. "This," as we are told by Gahn, "must be free from all impurity, and especially from sulphuric acid, as this will be decomposed, and sulphuret of soda will be formed, which will dissolve the metals we wish to reduce,

G 2

and produce a bead of coloured glass with substances that would otherwise give a colourless one." But the barilla, or soda of commerce (which is a sub-carbonate) is almost always in a state of impurity. Frequently it contains sulphate of soda and carbonate of potass, and is scarcely ever found free from muriate of soda. Klaproth's mode of obtaining this salt perfectly pure, is as follows:—Dissolve common sub-carbonate of soda in water, and saturate this solution with nitric acid, taking care that the acid is a little in excess. Then separate the sulphuric acid by nitrate of barytes, and the muriatic acid by nitrate of silver. The fluid thus purified, which is nitrate of soda, must be evaporated to dryness, fused, and decomposed by detonation with charcoal. The residue, is to be lixiviated, and the solution to be set to crystallize, when the sub-carbonate of soda will be produced. If the salt be adulterated with potass, tartaric acid will form a precipitate in a pretty strong solution of it.—The following simple process will give sub-carbonate of soda in a state of sufficient purity for all purposes but those of great nicety. Dissolve the salt which is obtained from the druggists, in a small portion of water, and filter the solution; this clears it from the admixture of earthy bodies. Slowly evaporate the filtered solution by a slow heat; very small crystals of muriate of soda will form on the surface, and must be skimmed off. When these cease to appear, the solution may be suffered to cool, and the purified sub-carbonate of soda will crystallize.—The salt is to be reduced to fine powder for use. If we ignite it previously, we get rid of its water of crystallization, and render it less bulky; but then, there is this inconvenience produced: When we touch it with a wet knife, the moisture gradually communicates to the whole mass, and causes it to cohere in coarse lumps.

160. *Of the Fusion of Bodies with Soda.*—When an assay is exposed to the blowpipe mixed with this

flux, it commonly happens that the soda sinks into the charcoal the instant it is melted. "Soda," says Gahn, "will not form a bead upon charcoal, but with a certain degree of heat will be absorbed. When, therefore, a substance is to be fused with soda, this flux must be added in very small portions, and a very moderate heat used at first, by which means a combination will take place, and the soda will not be absorbed." But if too large a quantity of soda has been added at first, and it has consequently been absorbed, this does not prevent its action on the assay; for if it be fusible with the soda, and the heat be continued, the flux is pumped up again, the assay effervesces on the surface, its edges are gradually rounded, a combination ensues, and the whole is transformed into a fluid glassy globule. An assay which may be decomposed by soda, but is not capable of fusion therewith, may be observed to swell up slowly and change its appearance without fusion. Some minerals are fusible with soda, in *powder*, but not in any other form; others, combine readily with very small portions of soda, melt with difficulty if more be added, and are absolutely infusible with a larger quantity. Again, "If we take too little soda," says Berzelius, "a part of the matter remains solid, and the rest forms a transparent vitreous covering around it; if we take too much, the glassy globule becomes opaque on cooling." From these examples, we see the necessity, ere we pronounce on the infusibility of any substance with soda before the blowpipe, to try a mixture of the mineral and flux, in different proportions.—This flux is particularly adapted to the solution of siliceous minerals.

161. Sometimes the glass formed with soda, acquires, as it cools, a deep yellow or hyacinth red colour. When this happens, the mineral or the soda contains sulphur or sulphuric acid. If it takes place with every glass which is formed by the same

soda, it is a proof that the flux contains sulphate of soda, in which case it must be rejected. But if the discolouration takes place only in particular instances, it is then the assay that contains sulphur or sulphuric acid.

162. *Reduction of Metallic Oxides by means of Soda.* —This subject is treated of by Gahn, in the article referred to at paragraph 9, and by Berzelius, in his work on the blowpipe. The following information is derived from these two sources.

163. If a small portion of oxide of tin be exposed to the blowpipe, on charcoal, a skilful operator may produce from it a small grain of metallic tin; but if a little soda be added, the reduction is easily effected, and so completely, that, if the oxide be pure, it is wholly converted into reguline tin. Soda, therefore, manifestly favours the reduction; although we are ignorant of its precise mode of action. But if the metallic oxide contain a foreign body, incapable of being reduced, the reduction of the former often becomes in consequence more difficult. Hence arises a question of considerable importance,—How are we to discover that an assay, consisting chiefly of substances that are irreducible, contains a reducible metal? and if it does, how can we prove it?—The following simple solution of this problem, was discovered by Gahn.

164. The glass bead, formed after the manner already pointed out, is to be kept in a state of fusion on the charcoal as long as it remains on the surface, and is not absorbed; in order that the metallic particles may be collected into a globule. A little more soda is then to be added, and the blast renewed and continued till the whole mass is absorbed by the charcoal. During this operation, as strong a heat as possible must be produced, by directing the reducing flame, so as entirely to cover the surface of the fused mass. The spot where the absorption takes place, is to be strongly ignited by a tube with a small aperture.

By this means, the portion of metal which previously escaped reduction, will now be brought to a metallic state. This done, the charcoal is to be extinguished by a drop or two of cold water, and that part containing the absorbed matter is to be taken out with a knife, and ground to a very fine powder, with distilled water, in a crystal or agate mortar. This powder is then to be washed with water to free it from the soda, which will be dissolved, and from the charcoal, which, forming its lightest part, will float on the water, and may be poured off. The dropping tube is made use of for the washing. In this process, we must particularly attend to three things. First, to take every particle of the assay out of the charcoal, and be cautious that nothing is lost in conveying it to the mortar. Secondly, to grind the carbonaceous mass to an impalpable powder. Thirdly, to decant the water, loaded with supernatant charcoal, with the greatest care and gentleness; so that the heavier metallic particles be not disturbed and carried off. The grinding and washing are to be repeated till every atom of the charcoal is got rid off; since the smallest remaining portion would confuse the examination of the assay, either by concealing the little metallic spangles, or by exhibiting itself a pseudo-metallic lustre sufficient to deceive an inexperienced eye.

165. When the whole of the charcoal is removed, the substance which remains behind may be examined. We must not be content with looking at the result with the naked eye, but must inspect it carefully by the microscope, and examine its relations both to reflected and transmitted light. If the assay contained no metallic substance, nothing will remain in the mortar after the last washing. But if it contained a portion of reducible metal, be it ever so small, it will certainly be found at the bottom of the mortar—if fusible and malleable, in the state of little brilliant leaves; if infusible and brittle, in the state of powder. In either case, the bottom of the

mortar will exhibit metalliferous traces. The flattening of malleable metals transforms an imperceptible particle into a round shining disc of sensible diameter. Thus we may detect, by the blowpipe, in an assay of the common magnitude, a portion of tin forming but the two hundredth part of its bulk. Of copper, even the slightest traces may be discovered. Most of the metals may be reduced by this process.

166. If an assay contains 8 or 10 per cent. of metal, its reduction is always effected at the first trial; but as its proportion decreases, the operation necessarily becomes more difficult. The learner should practise himself in this kind of process, by selecting a substance containing copper, and making several experiments of reduction with it; mixing it every time with a larger proportion of matter containing no copper: the metallic standard thus diminishes with every fresh experiment, and when such a proportion is arrived at, that the reduction of the copper is not effected by a first operation, the experiment must be repeated till it is successful. By exercising himself in different trials of this sort, he will soon become expert in such operations, which only require a little practice and address.

MIC. SALT.

167. Also called Microcosmic Salt, Salt of Phosphorus, and Phosphate of Soda and Ammonia.—Dr Ure tells us, that this compound is best procured by mixing equal parts of the phosphate of soda and phosphate of ammonia in solution, and then crystallising. A faint excess of ammonia is useful in the solution. Berzelius instructs us as follows: To procure this salt, dissolve 16 parts of sal-ammoniac in a very small quantity of boiling water, and add 100 parts of crystallized phosphate of soda; liquefy the whole together by heat, and filter the mixture while boiling hot; the double salt will crystallize as it cools. If mic. salt fused

before the blowpipe be not pure, it produces a glass which becomes opaque on cooling. In which case, it must be re-dissolved in a very small portion of boiling water, and crystallised afresh.

168. Mic. Salt may be kept for use, either in powder, or in crystals, which are generally of a convenient size for experiment. Exposed to the flame of the blowpipe on charcoal, it intumesces considerably, and continues boiling, with a crackling noise, until the water and ammonia have flown off. It is now converted into acid phosphate of soda, which fuses without agitation, and produces a pellucid spherule. Its use as a re-agent rests on the action of its free phosphoric acid, which seizes with avidity on the bases of the assays, and forms with them double salts; all of which are more or less fusible, and give useful indications, with respect to their nature, in the colour, and the degree of transparency which they possess. This salt, therefore, exhibits the action of acids on the assays, and is more particularly applicable to the examination of the metallic oxides, whose characteristic colours it developes much better than borax. Moreover, the clear globule which is formed, when the flux is first melted, continues soft much longer than that formed by borax, and therefore is more fit for the addition of the matter which is to be dissolved.

169. Having given a pretty-detailed account of the fluxes which are most generally used, we can but briefly notice the remaining substances. Indeed it is unnecessary to speak of them at any great length, as the use of most of them is very limited.

170. *Solution of Cobalt.*—This is nitrate of cobalt dissolved in water, and somewhat concentrated. It must be perfectly pure; and especially free from alkali. It is a test for alumina and magnesia; producing with the former, after violent ignition, a fine blue colour, with the latter, a pale rose red. This

re-agent operates even in the presence of silica.—
An assay which is absorbent is moistened with the
solution, and heated strongly, but not fused: if it
contains alumina, it becomes blue, if it contains
magnesia, it becomes red.   Mark, that, the assay
is not to be fused; only heated strongly.   The
reason of this is, that minerals which contain lime
or alkali, without alumina, also become blue with
solution of cobalt, when they are fused, but not till
then; whereas, alumina produces a blue colour by
heat alone, without fusion.—Hard stones are ground
with water in the agate mortar till reduced to a pap.
A drop of which is laid on the charcoal, and a drop
of the solution of cobalt is added to it; the mixture
is then exposed to the blowpipe, and gradually heated
to redness.   If the mass, when baked to a cake,
loosens from the charcoal, it may be held by the forceps.   When it has been heated bright red, it must
be left till it is completely cold, and the colour must
be examined by day-light.   We owe this test to the
experiments of Gahn.—It is proper to remark, that
the presence of a metal in the assay, or, of an alkali in
the solution, destroys the action of the test.

171. *Saltpetre.*—This is used to complete the oxidation of substances which have partly resisted the
action of the exterior flame.   The point of a long
thin crystal is plunged into an assay while it is hot
and liquid.

172. *Vitrified Boracic Acid.*—Kept in coarse powder.   Used to detect phosphoric acid in minerals.

173. *Gypsum* and *Fluor-spar.*—Kept well dried.
Used only to detect each other.   A piece of the
former with a rather smaller piece of the latter, fuse
together before the blowpipe into a transparent colourless bead, which assumes the appearance of an enamel when cold.   In other proportions they fuse, but
not perfectly.

174. *Tin,* "Employed in the state of foil, cut
into slips, and closely rolled up.   Its use is to pro-

mote reduction in the highest degree in the fused vitreous compounds." *Berzelius.* When the end of the roll is dipped into a fused globule, the tin acts by absorbing oxygen from the assay.

175. *Iron,* in the state of harpsichord wire, is employed to precipitate copper, lead, nickel, and antimony, and to separate them from sulphur or fixed acids. It is dipped into a fused assay, and the blowpipe directed upon it, when it becomes covered with the reduced metal.

176. *Silica,* in a state of purity, forms a fusible glass with soda, which serves to detect the presence of sulphur or sulphuric acid.

177. *Oxide of Copper* is used to detect the presence of muriatic acid.

178. *Test Papers,* respectively tinged with tincture of litmus, brazil wood, and turmeric, are useful reagents in a variety of cases.

---

149. We have now described the whole of the fluxes which are generally used in analyses by the blowpipe, and have detailed at length their individual characteristics. We shall complete this part of the subject by introducing *a tabular view of the phenomena which minerals, when heated with fluxes, on charcoal, before the blowpipe, usually develope.* The analyst will probably find this to be of considerable service; as it will show him, at one view, what he has principally to attend to, when he takes up the instrument to operate.

180. A MINERAL, ON CHARCOAL, WITH EITHER OF THE FLUXES, EXPOSED TO THE BLOWPIPE FLAME,

May, or may not,

A—*Fuse,*

   i. Wholly,
   ii. Partially,
   iii. Quickly,
   iv. Slowly,
   v. Assuming a thick pasty appearance,
   vi. Assuming the appearance of a liquid.

> REMARK.—When an assay only partly fuses, notice the appearance of that portion which remains uncombined, and what becomes of it. Sometimes, an assay is merely corroded on the surface; in other cases, it is divided into powder. The residue of some minerals which partially fuse with fluxes, is a sort of skeleton of silica. This is especially the case when siliceous minerals are fused with mic. salt: a glass is formed, which on cooling is sometimes opaque, sometimes transparent, and a silica skeleton remains unfused.

B—*Intumesce,*

   i. Slightly,
   ii. Violently.

C—*Effervesce,*

   i. Slightly,
   ii. Violently.

> REMARK.—Observe whether the Intumescence or Effervescence occurs before fusion or after.

D—*Become more Transparent,* or
Become less Transparent.

> REMARK.—The degree of the alteration in the transparency of a mineral, and the period of the process when the phenomena occurs, must be noticed. Some minerals before the blowpipe, change from opaque to transparent, others from transparent to opaque; while a considerable number preserve their original character, whether of opacity or transparency.

E—*Change the Colour of the Flux.*

> REMARKS.—1. Very remarkable changes in the colour of fluxes is produced by heating them with different minerals. Indeed, this is one of the most important of the pyrognostic characters. Paragraph 146, exhibits a curious and instructive example of the mutual action of borax and a metallic oxide. 2. Not only the colour acquired by a flux, and the degree of its intensity, but the period of the process at which the change takes place must be noticed. 3. When a change takes place more than once, every successive tinge must be particularized, and it must be noticed whether the alteration occurs in the exterior flame or the interior. 4. Sometimes the colour of the fused assay is so intense that the glass appears opaque. Gahn recommends, in this case, that the glass be broken, and a part of it mixed and fused with more of the flux, until the colour becomes more pure and distinct. Berzelius says, we may ascertain its transparency by holding the globule, in a certain direction, opposite the flame of the lamp, when we may distin-

guish, even in broad day-light, the reversed image of the flame painted on the glass, in the colour of the latter. The bead while soft may be flattened between the claws of the instrument represented by figure 32, previously heated; or it may be drawn out into a fine thread. 5. As the colour produced, is one of the most certain indications of the nature and proportions of the metallic matter contained in an assay, this point must be well attended to.

F—*Detonate.*

REMARK.—A slight Detonation sometimes takes place, on charcoal, before the blow-pipe, especially when Nitrates are operated upon.

G—*Be Absorbed by the Support.*

REMARK.—This generally happens when soda is used, at the early part of the process. See paragraph 160, where the method of treating minerals with soda is detailed.—Sometimes the liquid glass formed by the fused flux and assay is absorbed by the charcoal, leaving the metallic portion of the substance under examination, upon its surface, *reduced.* The method of separating the assay from the charcoal when absorbed, especially in experiments of reduction, is minutely detailed in paragraphs 162-6.

H—*Impart a Colour to the Flame.*

REMARK.—Sometimes an assay gives a colour to the flame when heated with one flux, but does not do so with another. The difference in such a case is to be noticed. The property of giving a colour to flame is characteristic of some minerals. See Barytes and Strontian in the experiments which follow.

I—*Fuse, but will yield a Result,*

Which may be

i. A Bead of Metal.

REMARK.—The nature of the metal is to be judged of from the individual characters of the metals. The method of separating very minute particles of metal from particles of flux and charcoal is detailed at paragraph 164.

ii. Ashes or Powder,
iii. A Glassy Globule—
  a Transparent, wholly, partially, or not at all,
  b Filled more or less with air bubbles,
  c Perfectly colourless,
  d Tinged to a greater or less degree with some colour,
  e Homogeneous,
  f Heterogeneous.

REMARKS.—1. It is to be noticed, whether a glass which appears dull be really opaque, or is rendered apparently so by the air bubbles it contains, or by the diffusion of reduced particles of metal through its mass. —2. It is also to be noticed, whether the appearance which a glassy globule has *when hot,* continues *during,* and *after cooling;* or, if any and what change takes place.—3. Another thing to be observed is, whether a substance which has once undergone fusion, continues fusible, with or without more flux, or has become incapable of it.—4. The degree of the transparency of a glass frequently depends

H 2

upon the relative proportions of the assay
and the flux. Each of the fluxes can only
dissolve and combine with a portion of a
mineral, proportionate to its own bulk.
Soda is very easily affected by a different
quantity. See the account of that flux.—
5. When borax is the flux used, it is of
importance to know what effect Flaming
(136.) has on the globule.

iv. A White Enamel,
   a Smooth,
   b Having the appearance of a frit,
   c Homogeneous,
   d Heterogeneous.

General Observations.—1. In all the preceding
cases, it is necessary to notice, whether the phenomena are produced by the *interior* or the *exterior*
flame; and whether upon the immediate application
of the flame, or after having for some time resisted
its action. It is also necessary to notice *the order*
in which the different phenomena are developed;
which will rarely be the same as that in which they
are here described. Often, indeed, several characteristics are exhibited at the same time, and it requires a quick eye, and close attention, to observe
and distinguish them.—2. As the nature of the products vary greatly, as may easily be imagined,
according to the properties of the fluxes which enter
into combination with them, particular observation
must be made of the appearances produced in an assay by different fluxes, with each of which a separate
piece of the mineral must be tried.—3. Be cautious
in the use of Fluxes, of suspending the blast too
soon, as their action upon some minerals is so slow
and gradual that to produce the full effect, they often
take some time. Do not therefore speedily get out of
patience, and erroneously declare a substance to be
infusible, which has not been fairly tried.—4. Ob-

serve the variety of the phenomena produced by adding the flux in small successive portions; the additions to continue till the assay is fully saturated. By proceeding thus, phenomena are often produced which would not otherwise be made manifest.—5. By an habitual use of the blowpipe in the examination of minerals, a number of fluxes might perhaps be found out, capable of producing effects different from these already in use; whereby distinct characters might be discovered for these mineral bodies which now either show ambiguous ones, or can scarcely be tried at all with the blowpipe.—6. An abundance of examples, illustrative of the above phenomena, are exhibited by the mineral characteristics in the following pages.

———

The General Principles explained in the preceding pages, shall now be applied to exemplify the use of the blowpipe in Chemistry and Mineralogy.

A SUMMARY VIEW OF THE SIMPLE BODIES, AND PRINCIPAL CHEMICAL COMPOUNDS, WHICH ENTER INTO THE COMPOSITION OF MINERALS, AND OF THE PHENOMENA WHICH THEY PRESENT, WHEN EXPOSED TO THE ACTION OF THE BLOWPIPE.

---

181. Before the student proceeds to the examination of minerals, he must become acquainted with the nature and habitudes of mineral constituents, otherwise he will be continually perplexed by the developement of the various phenomena. It is intended, therefore, to furnish him, in the present division of this work, with the necessary information on this point. In the first place, we shall give a short account of the Simple or Elementary Bodies, and of the most important Chemical Compounds, which enter into the composition of Minerals. We shall then state the relative proportions in which these occur in the Mineral Kingdom. And, lastly, we shall describe the Phenomena which they present, when exposed to the action of the Blowpipe. But, let us not be misunderstood. It is not our purpose to investigate the general chemical properties of those bodies: with those the student is sup-

posed to be already acquainted. What we intend to do, is, to shew in what state, in what abundance, and in what minerals they are found; and to describe the methods by which they are most easily detected and most accurately discriminated. A clear and comprehensive view will be thus afforded of the relation that Mineralogy bears to Chemistry; and the student will be furnished with a mass of information that will prove of the utmost utility when he enters upon the examination of minerals.

182. TABLE OF SIMPLE BODIES.

CLASS FIRST.

Bodies having an immense affinity for the simple bodies of the succeeding two classes; with which bodies they combine, and thereby form substances that are totally different in their properties from the substances of which they are composed:—

1. Oxygen	3. Iodine
2. Chlorine	4. Fluorine.

CLASS SECOND.

Bodies of a non-metallic nature, but inflammable or acidifiable:—

5. Hydrogen } *Gaseous Bodies.*
6. Nitrogen }
7. Carbon } *Fixed and Infusible Solids.*
8. Boron }
9. Sulphur } *Fusible and Volatile Solids.*
10. Phosphorus }

CLASS THIRD.

Inflammable substances of a metallic nature:—

This is the most numerous class of Simple Bodies; the individuals of which it is composed being in number *forty-three.* They combine with nearly all the ten bodies named above; but the most important compounds into which they enter, are the substances formed by their combination with oxygen.—The following list exhibits the metals arranged in the order of their perfectibility. Those at the head of the list are scarcely at all affected by the power of oxygen. But as we progressively descend, the influence of that beautiful element progressively increases ; so that, among those near the bottom, it exercises an almost despotic sway, which the magic pile of Volta, directed by the genius of Davy, can only suspend for a season. The emancipated metal soon relapses under the dominion of oxygen. —In an annexed column is detailed the nature of the products formed by a combination with oxygen.

11. Platinum		34. Osmium	
12. Gold		35. Rhodium	
13. Silver		36. Iridium	*Metallic*
14. Palladium		37. Uranium	*Oxides,*
15. Mercury		38. Titanium	*not much* *known.*
16. Copper	*Ordinary*	39. Cerium	
17. Iron	*Metallic*	40. Wodanium	
18. Tin	*Oxides, or* *Neutral*		
19. Lead	*Saliftable*	41. Potassium	
20. Nickel	*Bases.*	42. Sodium	*Alkalies.*
21. Cadmium		43. Lithium	
22. Zinc			
23. Bismuth			
24. Antimony		44. Calcium	
25. Manganese		45. Barium	
26. Cobalt		46. Strontium	
		47. Magnesium	
27. Tellurium		48. Yttrium	*Earths.*
28. Arsenic		49. Glucinum	
29. Chromium		50. Aluminum	
30. Molybdenum	*Acids.*	51. Thorinum	
31. Tungstenum		52. Zirconium	
32. Columbium		53. Silicium	
33. Selenium			

183. All the bodies enumerated above, with the single exception of Iodine, enter into the composition of minerals. But it is necessary to observe, that the *proportions* in which they do so are remarkably different ; as some of them occur in the greatest abundance, and others are exceedingly rare. Another singular thing is, that the substances which are found in the largest quantities are not those which occur in a simple state. *Oxygen*, for instance, which constitutes about 50 per cent of the rocks which form the chief part of the crust of the globe, has never been obtained in a state of purity ; while palladium, an element of the greatest rarity, is found in no other state than its original one, that of a metal. However, it is seldom that *any* of the Elementary bodies occur in an uncombined state ; indeed, of all the above-named fifty-three, it is only the twelve following that ever do so, and even these but in very small quantities :—

Carbon—found pure only in the diamond.
Sulphur—found pure near volcanoes.
The metals Gold, Silver, Platinum, Palladium, Mercury, Copper, Iron, Antimony, Bismuth, Arsenic.

184. But although the simple bodies are generally found, in a state of combination one with another, the substances they compose are not so complex, as, from the number and the relative abundance of the elements, one might be led to suppose. Few mineral species contain more than five or six elementary constituent parts, and many species only two or three. Some of the elements have no affinity for each other, are never found together, and cannot by any possibility be made to combine. And in all other bodies, which are not mechanical mixtures but chemical combinations, the ingredients or elements of which they are composed, are always found in certain unvarying and definite proportions.

185. There is a certain class of chemical bodies called BINARY COMPOUNDS, which are substances composed each of two elementary bodies combined in proportions which are well known and invariable. The earths and alkalies, substances which enter so largely into the composition of minerals, are all binary compounds, each being composed of two elementary bodies combined together. Now in the analysis of a mineral, it is only necessary for the chemist to ascertain the nature and proportions of the binary bodies which compose it, to learn what would be the

precise result if the analysis was carried to completion.
For example,

186. If common gypsum or sulphate of lime is analysed,
its binary compounds are found to be lime, sulphuric acid,
and water, and these are farther found to exist in gypsum
in the following proportions:—

Lime	32:7
Sulphuric Acid	46.3,
Water	21.

100

Here we have three substances, which, although commonly
called the elements of gypsum, are not, accurately speaking, elementary bodies at all: lime consisting of oxygen and
calcium; sulphuric acid of oxygen and sulphur; and water
of oxygen and hydrogen: So that, were we to obtain the
real elementary parts of gypsum, we should have calcium,
sulphur, hydrogen, and sulphur. But it is unnecessary, and
indeed in many cases impossible, to carry the analysis of a
mineral to such an extremity. For all common purposes,
it is sufficient to know the binary products of a substance;
because, as the nature and proportions of the elements of
these are already known, the results of an ultimate analysis can be obtained by the easier process of calculation.

187. The Binary Compounds that most frequently occur
in the Mineral Kingdom are the following:—1. The
Earths (334), 2. The Fixed Alkalies (327), 3. Ordinary
Metallic Oxides (204, 306), 4. Alloys (205), 5. Acids
(258, 357), 6. Sulphurets (199—202), 7. Carburets (196).—
The whole of these are described in the paragraphs which
are referred to by the figures within the parentheses.

188. Besides the binary compounds above mentioned,
which are those of most frequent occurrence in the mineral
kingdom, there are others, that occur much seldomer, but
which it is necessary to be acquainted with. For instance,
ammonia, which is a compound of hydrogen and nitrogen, and
common salt, which in a dry state is composed of chlorine
and sodium, are both binary compounds occuring in minerals.

189. The substances which are composed of three
elementary bodies (and there are many such in the mineral
kingdom) are called TERNARY COMPOUNDS. This class
of bodies is a very extensive one, and comprises substances

of considerable importance. Most of the substances termed
SALTS are *ternary compounds*.

OXYGEN.

190. This substance is so abundant a principle in
many of the oldest and most bulky masses forming
the crust of the globe, that it should probably be
considered as the commonest and most plentiful of
the mineral elements. It is necessary to mention
but few examples in support of this opinion. Of
Silica, the most abundant of all mineral substances,
oxygen forms 50 per cent.; of alumina, 47; of lime,
28; of magnesia, 40; of potass, 17; of soda, 25; of
water, 89; to which it may be added, that it occurs
in various proportions as an ingredient of nearly all
the metallic ores. Carbonate of lime, which is alone
supposed to constitute about the one-eighth part of
the crust of the globe, contains oxygen as a constituent in the proportion of nearly one-half; while,
in the argillaceous and siliceous rocks, it is still
more abundant. It is easily obtained in the state of
gas for chemical purposes; but as it has never been
found in a state of complete separation from other
substances, it is chiefly as a constituent of several
important mineral compounds that it has been noticed here. See Oxides (204, 306), Acids (258, 357);
Earths (334), Alkalies (327).

CHLORINE.

191. Chlorine is another substance that has never
been found pure. With sodium, the metallic base
of soda, it forms rock salt. It is a constituent of
horn lead, horn silver, and horn mercury. With
hydrogen it forms muriatic acid. It is never sought
for in a separate state by the Mineralogist.

IODINE.

192. This substance, as has been already mentioned, has not yet been found in any mineral.

I

FLUORINE.

193. This has not been exhibited in a separate state. It forms, with hydrogen, fluoric acid, which is a constituent of fluor-spar and of several other minerals.

HYDROGEN.

194. This has never been obtained in a purer form than that of gas. It is a constituent of water, ammonia, fluoric acid, and muriatic acid; all of which enter into the composition of minerals. It combines with carbon, and forms coal and bitumen. With sulphur it forms sulphuretted hydrogen gas, which is a constituent of stinkstone. In some places, hydrogen gas issues in a state of greater or less purity, from crevices in the earth.

NITROGEN.

195. The only claims of this element to be considered a mineral constituent, are, that it enters into the composition of ammonia and nitric acid, both of which occur in minerals.

CARBON.

196. This occurs in a nearly pure state in graphite, and composes the entire substance of the diamond. It forms the basis of coal, bitumen, amber, and other combustibles; and is a constituent of a variety of other minerals. Its most important mineral character is, that it forms the base of the carbonic acid, which enters as an ingredient into the composition of all limestone rocks, in the proportion of 44 per cent. The binary compounds it forms with the metals and combustibles are called *carburets*.

197. Carbon is gradually consumed when ignited; producing neither flame, smell, nor smoke; pulverised and mixed with saltpetre, it detonates on platinum foil, and leaves a residuum of carbonate of

potass. When hydrogen is present, as in coal, it burns with flame and smoke.

BORON.

198. As a mineral constituent, this is only remarkable as being the base of the boracic acid, which enters into the composition of borax, the boracite, and the datholite. Boron never occurs pure, and can only be obtained from its binary compounds by powerful voltaic agency.

SULPHUR.

199. This occurs in a state of purity, and variously combined with other elements. With oxygen, composing sulphuric acid, it is widely distributed; forming, with metals, earths, and alkalies, both soluble and insoluble salts. The vitriols are soluble metallic sulphates; Glaubersalt is a soluble alkaline sulphate; Gypsum is an insoluble earthy sulphate. Sulphur forms binary compounds with many of the metals which are termed *sulphurets*. Iron pyrites and copper pyrites are both sulphurets.

200. The presence of sulphur in an assay is easily detected by the sulphurous smell it gives out when heated in the matrass or on charcoal. When it occurs in very small portions, it is discovered by other means. See sulphuric acid.

201. *Metallic Sulphurets.*—These are known by the odour of sulphurous acid which they emit when heated. When a metallic sulphuret is assayed, the object in view is to learn what metal the sulphur is combined with. We first drive off the sulphur by roasting; which is performed either in a glass tube or on charcoal. (See 131.) It cannot be done on platinum foil. An assay for roasting should be chosen in the form of a thin lamina, which is the figure best adapted to the operation. Care must be taken not to melt the assay while roasting it. If fusion happen to takes place, a fresh assay must be

operated upon. When the roasting is completed,
and the sulphur entirely expelled, we may then
avail ourselves of the use of the fluxes, with which
nothing can be done until then. The reduction by
means of soda particularly requires the complete ex-
pulsion of sulphur from the assay.

202. The following additional remarks on this subject,
by Engeström, will be found useful :—Whenever an ore is
to be tried, a small bit is to be laid on the charcoal,
and the flame blown on it slowly : upon which the
sulphur, and arsenic if it contains any, will rise from it
in the form of smoke. These two volatile substances
are easily distinguished from one another by the smell ;
that arising from sulphur being sufficiently well-known, that
of arsenic resembling the odour of garlic. The flame ought
to be blown very gently, as long as any smoke is seen to
part from the ore, but, after that, the heat must be aug-
mented by degrees, in order to make the roasting as perfect
as possible. If the heat is applied very strongly from the
beginning upon an ore that contains much sulphur or arsenic, the ore will presently melt, and yet lose very little of
its volatile ingredients ; by which means the roasting is
rendered extremely imperfect.—The ore being properly
roasted, the metals contained in it may be discovered,
either by being melted alone, or with fluxes ; on which they
present themselves, either in their pure metallic state, or by
tinging the slag with colours peculiar to each of them. In
these experiments, it is not to be expected that the quantity
of metal contained in the ore should be exactly determined :
this must be done in larger laboratories. This cannot, however, be looked upon as any defect, since it is sufficient for
a mineralogist, in many cases, to know merely the sort
of metal that is contained in an ore. But in the blowpipe
processes, a practised eye can readily distinguish the phenomena presented by a large proportion of metal from those
presented by a very small proportion.

PHOSPHORUS.

203. This is never found native ; but it occurs, in
combination with oxygen, in phosphoric acid, which
combines with the metals lead, manganese, copper, and uranium, in various proportions. Phosphoric acid, is also found combined with the earth lime.

In all these cases, it forms salts which are termed phosphates.

___

204. In treating of the metals, it will be most convenient to give the habitudes of the oxides along with those of the pure metals ; which is the plan we shall adopt. The oxides of the first division of metals, or *ordinary metallic oxides*, as they may be termed, require few general remarks. They are distinguished from the metallic acids, the earths, and alkalies, by not possessing the peculiar properties by which those substances are characterised ; and by being reduced, with far greater facility, than they are, to the metallic state. This remark, however, does not hold good with respect to every metal, as the reader will perceive upon examining the following characters.

205. The method of treating *alloys*, will be described when speaking of the metals which usually occur alloyed. Occasionally a few observations will be made as to the phenomena which characterise the *ores* of the different metals. A number of experiments likewise will be described, for the purpose of exhibiting the manner of proceeding in cases where the young operator might find himself at a loss.

PLATINUM.

206. This metal is found only in the native state and but in small quantities. It is generally alloyed with palladium, rhodium, iridium, and osmium ; and it often contains a minute portion of iron or chromium.

207. *Platinum in the native state, alone,* before the blowpipe, is infusible.

208. With *fluxes,* it is infusible, and incapable of oxidation ; but the fluxes generally acquire a colour from the foreign bodies by which it is contaminated.

209. Berzelius informs us, that the best method of examining native platinum, is by cupellation, with perfectly pure lead, on bone ashes (as described at 132). In this experiment we can judge both of the presence and nature of the foreign metals, by the colour of the cupel loaded with the oxide of lead. At the end of the cupellation, the platinum is ob-

tained in the form of a greyish coloured infusible
malleable mass.

210. *Oxide of Platinum,* (obtained by precipitation
from a solution of nitro-muriate of platinum, by
means of potass or ammonia,) on charcoal, with *mic.
salt,* is reduced to the metallic state ; yielding a
small malleable globule.

GOLD.

211. Few metals have a greater number of localities than this, yet, as it only occurs in small quantities at each locality, it is by no means an abundant
substance. It is generally found in the metallic
form, but is in most cases alloyed with small portions of silver, copper, or some other metal.

212. *Gold in the metallic state, alone,* on charcoal,
melts into a globule, but cannot be oxidated.

213. The *fluxes* have no action upon it, and can
only serve to discover the foreign metals that are
alloyed with it.

214. Like platinum, it is best examined by cupellation, which yields a button of pure gold.

SILVER.

215. The ores of this metal are numerous. It
occurs native ; also, combined with antimony, iron,
arsenic, lead, copper, and bismuth ; and mineralized
by sulphur, by the carbonic and sulphuric acids,
and by chlorine.

216. *Silver in the metallic state, alone,* easily
fuses ; forming a beautiful white globule, which
resists oxidation.

217. With *mic. salt,* in the exterior flame, produces a glass having the colours of opal. Viewed
by refraction, it is yellow by day light, red by candle
light. In the interior flame it acquires a grey
colour.

218. *Oxide of Silver, alone,* on platinum or charcoal, is instantly reduced.

219. With *borax*, partly reduces, partly dissolves; the glass formed by the exterior flame, on cooling becomes milky, that formed by the interior flame has a greyish appearance, arising from the particles of reduced metal dispersed through it.— With *mic. salt*, dissolves quickly and copiously, but upon cooling grows opaque and of a whitish yellow. If copper be present, it is discovered by a green colour. If gold be present, a ruby tinge is produced. The globules appear opaline and can scarcely be got pellucid, except the quantity of oxide be very small. The globule loaded with dissolved silver, during fusion in the spoon, covers a piece of copper added to it, with silver, and becomes itself of a pellucid green. Antimony speedily takes away the milky opacity of dissolved luna cornea, and separates the silver in distinct grains. Cobalt and some other metals also precipitate silver in the same way.

PALLADIUM.

220. A very rare metal. It occurs native; also, alloyed with platinum.

221. Like platinum, this metal is best examined, before the blowpipe, by cupellation.

MERCURY.

222. The ores of mercury are not numerous, but some of them are very rich. The metal is not a scarce one. It occurs native; also, alloyed with silver, and combined with sulphur and certain acids.

223. All the compounds of mercury are volatile, and therefore not acted upon by fluxes. Mercurial minerals are assayed, by mixing them with oxide of lead, or iron filings, or metallic tin, and heating the mixture in a small matrass or glass tube; upon which, the mercury sublimes in the metallic state, and adheres to the cold part of the glass, in the form of a grey powder which agitation brings together

into fluid globules. The fixed heterogeneous matters remain behind.

224. *Oxide of Mercury* becomes black, and if pure, is entirely volatilised; this is therefore a test for adulterations.—The *fluxes* take it up with effervescence, but it is soon driven off.

225. Mercury combined with sulphur liquefies upon the charcoal, produces a blue flame, smokes, and gradually disappears. But if cinnabar be exposed to the blowpipe, upon polished copper, the mercurial globules are fixed all round.

COPPER.

226. This is a very abundant metal. Its ores are numerous. It occurs in the native state; also, combined with iron, antimony, silver and arsenic; and mineralized by oxygen, sulphur, and the carbonic, muriatic, phosphoric, and arsenic acids.

227. *Oxide of Copper, alone,* is not altered in the oxidating flame, but becomes *protoxide* in the reducing flame; with a strong heat, it fuses into a metallic globule.

228. Both with *mic. salt* and *borax* it forms a pellucid glass, which is yellow-green while hot, but blue-green when cold. If strongly heated, on charcoal, in the interior flame, the colour disappears, and the metal is reduced. If the quantity of oxide is so small that the green colour is not perceptible, its presence may be detected by the addition of a little tin, which occasions a reduction of the oxide to protoxide, and produces an opaque red glass. If the oxide has been fused with borax, this colour is longer preserved; but if with mic. salt, it soon disappears by a continuance of heat. The copper may also be precipitated upon iron, but the glass must be first saturated with iron. Vestiges of copper so faint as scarcely to tinge the flux, precipitate a visible pellicle upon a piece of polished iron added to it during strong fusion, and the globule takes the colour of

iron: in this way the smallest portions of copper may be discovered. Alkalies or lime promote this precipitation. If the glass containing copper be exposed to a smoky flame, the copper is superficially reduced, and the glass covered while hot with an iridescent pellicle, which is not always permanent after cooling. The globule made green by copper, fused in the spoon with a small piece of tin, until the colour is discharged, yields a spherule of tin mixed with copper, hard and brittle. In this case, the precipitated metal pervades the whole of the mass, and does not adhere to the surface. Cobalt precipitates the oxide of copper dissolved in the spoon with a flux, in a metallic form, and imparts its own colour to the glass, which nickel cannot do. Zinc also precipitates it separately, and rarely upon its own surface, as its fusion can scarcely be avoided.

229. With *soda*, on the platinum wire, it melts and gives a glass, fine green colour while hot, colourless and opaque when cold. On charcoal, by the reducing process, metallic copper is very easily obtainable.— The blowpipe is capable of detecting copper in minerals, when occuring in proportions so minute as to elude the search in all other processes; except, indeed, it be combined with other easily reducible metals, which are liable to disguise its properties. When this is the case, borax and tin are to be employed to effect reduction.—When copper and iron occur together, the same operation separately reduces each metal into distinct metallic particles, which may be distinguished by their particular colours and properties, or separated by the magnet.

230. The *Salts of Copper* tinge the flame of the blowpipe. Both sulphate and nitrate of copper produce a greenness; but muriate of copper acts with far greater efficacy. The green crystals of this salt first grow red by the exterior flame; after which, they liquefy and become black. The flame meanwhile is first tinged deep blue, which soon verges to green.

When thus coloured, the flame expands considerably, and remains so until the whole mass of the salt is dissipated. This green salt, added to *mic. salt* in fusion, immediately produces a beautiful flame; the clear globule is tinged green, and does not grow opaque or brown, unless a large quantity of the mic. salt be added; whereas, it takes place much sooner upon the addition of a small quantity of borax.

231. *Mineralized Copper.* When pyrites contain copper, though the proportion be less than the hundredth part of their weight, yet its presence may be detected by these experiments in small: Let a grain, the size of a flax seed, be roasted, but not so much as to expel all the sulphur; let it then be well dissolved by borax, a polished rod of iron added, and the fusion continued, until the surface, when cooled, loses all splendour. As much borax is required as is sufficient to make the whole as large as a grain of hemp seed. Slowness of fusion is injurious, and by too great tenuity the precipitation is retarded; this may be corrected by the addition of a little lime. Too much calcination is inconvenient; for by this, the globule forms slowly, is somewhat spread, becomes knotty when warm, corrodes the charcoal, destroys the iron, and the copper does not precipitate distinctly;—this defect is amended by a small portion of the crude ore.—When the globule is properly fused as directed, immediately upon stopping the flame, let it be thrown into cold water, in order that it may break suddenly. If the cupreous contents of the ore be less than the hundredth part of its bulk, one end of the wire only is covered with copper, which otherwise would be entirely covered.

232. The celebrated Gahn, who has examined copper ores with peculiar accuracy, has another method of discovering the smallest traces of that metal; namely, a grain of the ore, well freed from sulphur by calcination, is exposed to the action of the flame, driven suddenly upon it, *per vices*; and

at these instants a cupreous splendour appears upon the surface; which otherwise is black, and this splendour is the more quickly produced, in proportion as the ore is poorer.—Cupreous Pyrites, on roasting, tinge the flame green.

IRON.

233. This is an ingredient of almost every rock, from the oldest primitive to the newest alluvial; of a great number of earthy and metalliferous minerals; and of all soils: it is therefore considered to be the most abundant and most generally diffused of all the metals. In general, it is in the state of an oxide, except when combined with sulphur. It occurs in a state of combination with the oxides of titanium, manganese, and chromium; with the phosphoric, sulphuric, carbonic, muriatic, and arsenic acids; and with silica, alumina, lime, and water. What is called native iron, is an alloy of iron and nickel.

234. *Iron in the metallic state, alone*, is infusible, but is converted into an oxide.

235. With *borax*, it is infusible.—With *mic. salt*, it fuses, and forms a brittle mass.

236. *Oxide of Iron, alone*, on charcoal, in the oxidating flame, is not changed; in the reducing flame, it becomes black and magnetic. On platinum, in the reducing flame, it fuses.

237. With *borax*, in the oxidating flame, it produces a glass, dull blood red while hot, clear and yellow or colourless when cold. In the interior flame, on charcoal, it is reduced to protoxide, becoming of a green colour and attractible by the magnet. The protoxide, with borax, forms a green glass, which, by increasing the proportion of the assay, passes through bottle green to black and opaque. Berzelius says, that the bottle green glass, is formed by a mixture of peroxide and protoxide, and that, when the former is wholly reduced to the latter, the glass becomes perfectly transparent, possessing a green colour, which

is beautiful while hot, but fades on cooling. The reduction of iron from the peroxide to the protoxide is much facilitated by the addition of metallic tin.—With *mic. salt*, it behaves as with borax.—With *soda*, on charcoal, it does not dissolve; but sinks with the flux into the support; yields by reduction, a grey metallic magnetic powder.—When placed on the wick of a candle, it burns with the crackling noise peculiar to iron.

TIN.

238. This metal never occurs pure, but always in the state of an oxide, which is often combined with small portions of oxide of iron and silica. It is also found combined with sulphur and copper.

239. *Tin in the metallic state, alone*, melts at a very low heat, ignites, and is converted into oxide of tin. It is one of the most easily fusible metals. Berzelius gives us the following experiment :—If a red hot grain of metallic tin be thrown on a paper tray, it will divide into several smaller grains, which skip about the paper and burn with a very vivid light.

Curious example of the Combination of Tin with Platinum.—This is an experiment of Dr. Clarke's. If you take two pieces of tin-foil and platinum-foil of equal dimensions, and after rolling them together, place them upon charcoal, and direct the flame of a candle cautiously towards the edges of the roll,—at about a red heat, the two metals will combine with a sort of explosive force, scattering their melted particles off the charcoal, and emitting light and heat in a very surprising manner. Then there will remain upon the charcoal, a film of glass, which, by further urging the flame towards it, will melt into a highly transparent globule of a sapphire-blue colour. Also, if the platinum and tin be placed beside each other, as soon as the platinum becomes heated, you will observe a beautiful play of light upon the surface of the tin, which becomes highly iridescent before it melts.

240. *Oxide of Tin, Alone*, on platinum, in form of hydrate, and in its highest degree of purity, becomes yellow when heated, then red, and when ap-

proaching to ignition, black. If iron or lead be mixed with it, the colour is dark brown when heated. These colours become yellowish as the substance cools. On charcoal, in the interior flame, it becomes and continues white; and, if originally white and free from water, it undergoes no change of colour by heating. It is very easily reduced to the metallic state without addition, but the reduction is promoted by adding a drop of solution of soda or potass.

241. With *borax*, it melts sparingly and with much difficulty; and forms a glass which is rendered opaque neither by cooling nor flaming, unless it is supersaturated with oxide. With *mic. salt*, it behaves as with borax: the presence of arsenic, renders the glass opaque. With *soda*, on the platinum wire, it effervesces and forms an infusible mass. With soda, on charcoal, it is readily reduced into a globule of metallic tin, unless iron be present. If the proportion of iron be very small, its reduction may be prevented by the addition of borax to the soda. Tin may be detected, by the reducing process, even when it forms but the two hundredth part of the weight of an assay.—A very small quantity of tin dissolved in any flux, may be distinctly precipitated upon iron.

242. Tin may be melted out of the pure tin ores, in its metallic state. Some of these ores melt very easily, and yield their metal in quantity, if only exposed to the fire by themselves: but others are more refractory; and as these melt very slowly, the tin, which exhudes in the form of small globules, is inflamed and converted into oxide, before the globules can be collected into one sufficiently large to resist the immediate action of the fire, and to be perceived by the naked eye. It is necessary therefore to add a little borax to these ores, and to direct a strong flame upon the assay. The borax preserves the metal from being too soon calcined, and contributes to the readier collecting of the small metallic particles, which are soon seen to form themselves into a globule of metallic tin at the bottom of the whole mass, nearest to the charcoal. As soon as so much of the metallic tin is produced, as is sufficient to convince the operator of its presence, the fire ought to be discontinued;

K

although the whole of the ore may not be melted. The reason for which is this :—The whole of this kind of ore can be seldom or never reduced into metal by means of these experiments ; a great proportion being always calcined ; and if the fire is continued too long, perhaps even the metal already reduced may likewise be burnt : for the tin is very soon destroyed from its metallic state by the fire. —*Engeström.*

LEAD.

243. Next to iron, this may be considered as the most abundant and universally diffused of the metals. It is never found in the native state ; but its ores are very numerous, and appear under very different circumstances and aspects, and present a great variety of combinations. Mineralized by sulphur, it occurs in great abundance. It is also found in combination with oxygen, with the carbonic, muriatic, phosphoric, arsenic, molybdic, and chromic acids ; with the metals antimony, iron, manganese, and silver ; with the earths, silica, alumina, lime, and magnesia; and with water. Some of the ores contain several of these substances ; a few have a metallic aspect ; but others assume the appearance of earthy minerals.

244. *Lead, in the metallic state, alone,* is very easily fusible. The melted globule continues for some time to retain a metallic splendour ; but in a more intense heat, it boils and fumes ; hence oxide of zinc is deposited in a yellow circle upon the charcoal. It communicates a scarcely visible yellow colour to fluxes, and when the quantity is large, the globule, on cooling, assumes a milky opacity. When dissolved, it is not precipitated by copper. The metals do not precipitate it from sulphur in the same order as from acids.

245. *Oxide of Lead.*—*Alone,* the red oxide first becomes black, and is afterwards changed to *yellow* oxide, which melts into an orange-coloured glass ; the latter, on charcoal, is very quickly reduced, with effervescence, into a globule of metallic lead.

246. With *borax*, on the platinum wire, it melts with ease into a clear glass, which is yellow while hot, colourless when cold. On charcoal, it flows about in a liquid state, and reduces.—With *mic. salt*, it forms a clear colourless glass; but is not so easily reduced as with borax.—With *soda*, on charcoal, it is instantly reduced.

NICKEL.

247. This is a rare metal. It occurs as an oxide; also combined with arsenic and with iron.

248. *Pure Nickel* is not fusible by the blowpipe. Nickel containing arsenic does not fuse with *soda*; but if *borax* is added, it fuses into a globule which is malleable in a certain degree, and highly magnetic.

249. *Oxide of Nickel, alone,* at the extremity of the oxidating flame, becomes black; in the reducing flame, greenish grey.

250. *With borax,* it dissolves readily; the glass, while hot, is of a dirty dark red colour, but it becomes paler and yellowish as it cools. When this glass is long exposed to a high degree of heat in the reducing flame, the colour passes from reddish to blackish and opaque; then becomes blackish grey, and translucent; then paler reddish grey, and clearer; and, lastly, translucent; metallic nickel being precipitated in small white globules. The red colour seems here to be produced by the entire fusion or solution of the oxide; the black by incipient reduction; and the grey by the dispersion through the glass of the minute metallic particles, before they combine and form small globules. When a little soda is added to the glass formed with borax, the reduction is more easily effected, and the metal is collected into a single globule. When the oxide of nickel contains iron, the glass retains its own colour while hot, but assumes that of the iron as it cools; if cobalt is present, the colour of that metal prevails;

if arsenic, it melts into a bead. Nickel is precipitated either on iron or copper.—With *mic. salt*, it exhibits nearly the same phenomena as with borax, but the colour fades to a much greater degree on cooling. It behaves in the same way both in the exterior and interior flame. *Saltpetre* added to the bead, makes it froth, and it becomes first red-brown, then paler. With *mic. salt* and *tin*, the nickel precipitates, and the glass becomes colourless.—With *soda*, it does not dissolve, but, by the reducing process, nickel is obtained in the form of minute, whitish, splendent, metallic particles, which are very strongly magnetic.

CADMIUM.

251. A very rare metal. Occurs only in particular ores of zinc.

252. *Oxide of Cadmium, alone*, on charcoal, is dissipated in a few seconds; the support becoming covered with a red or orange-coloured powder. Zinc ores, and other substances containing oxide of cadmium, when heated on charcoal for a single instant, deposit an orange-coloured ring round the assay. On platinum foil, the orange ring is conspicuous, and by exposing it to the point of the blue flame, the metal is revived, and a deposition is formed upon the support, having the appearance of polished bronze or copper.

253. With *borax*, on the platinum wire, it fuses and gives a clear glass, yellow while hot, colourless when cold. If nearly saturated, it is rendered opaque by flaming. On charcoal, it bubbles constantly, is reduced, and sublimed; a yellow oxide covers the charcoal at the end of the experiment.—With *mic. salt*, it fuses; glass, clear while hot, opaque when cold.—With *soda*, on the platinum wire, it is infusible. On charcoal, it reduces, and is sublimed; leaving a yellow ring.

ZINC.

254. Never found in a pure metallic state. Generally occurs mineralized by sulphur, oxygen, and the carbonic or sulphuric acid; also combined with oxide of iron, silica, and water. Its ores, which are not very numerous, although the metal is by no means scarce, have commonly an earthy appearance.

255. *Zinc, in the metallic state, alone*, readily fuses and takes fire; burning with a beautiful blueish green flame, and sending forth clouds of white lanuginous oxide, by which the flame is speedily extinguished. If the reguline nucleus, included in this lanuginous matter, be urged by the fire, it is inflamed now and then, and flies about with slight explosions.

256. With *borax*, it ramifies, and at first tinges the flame; it continually diminishes, and the flux spreads on the charcoal.—With *mic. salt*, it ramifies, and sends forth flashes, with a crackling noise. By too great a degree of heat, it is exploded, upon which it throws out ignited particles.

257. *Oxide of Zinc, alone*, on charcoal, when heated, assumes a yellowish splendour, which, when the flame ceases, vanishes—and the assay whitens as it cools. It is infusible, but when ignited gives out a brilliant light, and in the interior flame evolves copious white fumes which condense on the charcoal in the form of wool.

258. With *borax*, it melts easily into a clear glass which turns milky if flamed, or with more oxide becomes enamel white on cooling. In the interior flame, the oxide is converted into white fumes which form clouds around the globule, and condense in a pulverulent form on the charcoal.—With *mic. salt* it behaves as with borax, but is rather less easily sublimed.—With *soda*, it does not melt, but in the interior flame is reduced, and burns with its characteristic flame, depositing its white oxide on the sup-

port. By this process, zinc may be easily detected, even in the automalite.—With *solution of cobalt*, dried and ignited, it becomes green.—Mixed with *oxide of copper*, and reduced, the zinc will be fixed, and brass be obtained.—But one of the most unequivocal characters of the oxide of zinc is obtained by dissolving it in vinegar, evaporating the solution to dryness, and exposing it to the flame of a lamp, when it burns with the peculiar flame of zinc.

BISMUTH.

259. The ores of bismuth are few. It occurs in the native state, somewhat alloyed with arsenic; also combined with sulphur, silver, and cobalt. It is not a common metal.

260. *Bismuth, in the metallic state, alone*, in the matrass, does not sublime at the heat which fuses glass; it is distinguished by this character, from tellurium. In the open tube, it is partly converted into an oxide, which is dull brown while hot, yellowish when cold. It gives scarcely any fumes; by this character it is distinguished both from antimony and tellurium. On platinum, it easily fuses. On charcoal, in a gentle heat, it flies off in fumes. The assay leaves a reddish mark on the support, which may be driven off by the reducing flame, to which it does not give a colour. The mark left by antimony, in similar circumstances, colours the flame greenish-blue; and that left by tellurium, deep green.

261. *Oxide of Bismuth, alone*, on platinum foil, melts readily into a brown glass, which becomes brighter as it cools. If violently heated, it reduces and perforates the support. On charcoal, it is very readily reduced into metallic globules.

262. With *borax*, it forms a grey glass, which cannot be freed from bubbles; this decrepitates in the interior flame, and the metal is reduced and volatilized.—With *mic. salt*, it forms a brownish-

yellow glass, which partly loses its transparency and
becomes pale, when cold. But if more oxide be
added, the glass becomes opaque. Bismuth is
easily precipitated by copper and iron.

ANTIMONY.

263. This is not a scarce metal; yet its ores are
but few in number. It occurs in the native state,
alloyed by small portions of silver and iron; also
combined with sulphur, silica, and oxygen.

264. *Antimony, in the metallic state, alone*, on
charcoal, easily melts, and when ignited by being
heated to redness, continues to burn for a considerable time after the action of the blowpipe is suspended. During the combustion, it gives off dense
and copious white fumes of oxide of antimony,
which rise perpendicularly, and afterwards condense on the metallic globule, and envelope it in a
net work of radiating acicular pearly crystals. This
affords a very beautiful object. In the open tube,
heated to redness, it burns slowly, depositing a white
oxide on the sides of the glass. In the matrass, it
does not sublime at the fusing point of glass.

265. *Oxide of Antimony, alone*, on charcoal, melts
easily and is reduced to the metallic state; colouring the flame greenish. On platinum, it melts
easily and yields a white vapour. Sometimes it
takes fire like tinder, without fusing, and produces
antimonious acid.

266. *Antimonious Acid, alone*, on charcoal, neither
melts nor reduces; but gives out a bright light, and
fumes.

267. *Antimonic Acid, alone*, on charcoal, instantly
turns white, and becomes antimonious acid; if it
contains water, it gives it off on ignition; at the
same time, changing colour from white to yellow,
and from yellow again to white.

268. The oxide and acids of antimony present
with the *Fluxes* the same phenomena.—With *borax*,

they dissolve largely; the glass is transparent, of a
yellow colour while hot, but nearly colourless when
cold; if supersaturated, the antimony sublimes and
condenses on the charcoal round the assay; the glass
when violently heated in the internal flame, becomes
opaque and greyish, in consequence of a partial reduction of the assay to the metallic state.—With
mic. salt, on the platinum wire, in the external flame,
they form a transparent yellowish glass; of which
the colour flies on cooling.—With *soda*, on charcoal,
they are converted into metallic antimony by the
reducing process.—Antimony dissolved in a flux may
be precipitated by iron and copper, but not by gold.

269. *Sulphuret of Antimony, alone*, on charcoal,
melts, smokes, spreads, and is absorbed; leaving nothing but a ring on the spot where it lay. In the
open tube, it sublimes; producing oxide of antimony
and a portion of antimonious acid, the latter of
which adheres to the surface of the glass, in the
form of a white coating, after the oxide has been
driven off by the heat.

270. *Alloys of Antimony*, when roasted in the
open tube, if combined with metals that are very oxid-
able, give out antimonious acid, the vapour of which
is infusible and fixed; but if combined with silver
or copper, they give out fumes of oxide of antimony,
which condense on the glass, but are volatile. A
pungent odour is produced.

MANGANESE.

271. This substance is very difficult to be procured in the metallic state, in consequence of the
strong affinity which it has for oxygen. It occurs
in the state of oxide, in small proportions, in a very
great number of minerals, both earthy and metallic.
It is diffused through all the three kingdoms of
nature.

272. *Oxide of Manganese, alone*, on charcoal, is
infusible. A strong heat changes its colour to brown.

273. With *borax*, on charcoal, in the exterior flame, it gives a clear glass of an amethyst colour; it becomes colourless in the interior flame. See paragraph 146.—With *mic. salt*, it behaves as with borax, but the colour is not so deep; while fusing in the oxidating flame, the glass boils up and disengages gas; in the reducing flame, it fuses quietly. When the manganese, from its combination with iron, or any other cause, does not produce a sufficiently intense colour in the glass, a little saltpetre may be added to it while in a state of fusion; the glass then becomes dark violet while hot, and reddish violet when cold.—With *soda*, on platinum foil, it forms a clear green glass, which becomes blueish green on cooling. This test is so extremely delicate, that a portion of manganese forming but the thousandth part of the assay, imparts a sensibly green colour to the flux. With soda, on charcoal, it is not reduced to the metallic state.—Manganese is one of those substances which often colour the glass so intensely that it appears opaque. See paragraph 180, letter E, remark 4.

COBALT.

274. This is one of the scarce metals. It is not found in the native state. In its ores, it is combined with iron, arsenic, and sulphur. Sometimes it is mineralized by sulphuric acid.

275. It is very difficult of fusion.

276. *Oxide of Cobalt, alone*, on charcoal, becomes black in the oxidating, and grey in the reducing flame. It does not fuse.

277. *With borax*, a small quantity melts easily into a transparent glass of a beautiful blue colour, which does not become opaque by flaming. By transmitted light, the glass is reddish, and by farther additions of the oxide it passes through dark blue to black. This colour is pertinaceous in the fire. That an opportunity may be obtained of seeing the colour

of the glass distinctly, the operator must take hold, with a pair of pincers, of a little of the glass, and draw it out, slowly in the beginning, but afterwards very quickly, before it cools ; whereby a thread of the coloured glass is procured, more or less thick, wherein the colour may be more easily seen, either by day light or candle light, than if it was left in a globular form. This thread melts easily if only put in the flame of a candle, without the help of the blowpipe. The metal may be precipitated from the dark blue glass, by inserting a *steel wire* into the mass while in fusion. It is not precipitated upon copper. It is malleable if the oxide has been free from arsenic, and may be collected by the magnet, for which it has a strong attraction. It is distinguished from iron by the absence of any crackling sound when placed on the wick of a candle.—*With mic. salt*, it behaves as with borax. By candle light, the pure blue glass appears violet, the pale blue appears rose coloured.—*With soda*, on platinum, it melts partially into a thin liquid ; the fused portion assuming a red colour, which on cooling changes to grey.—*With sub-carbonate of potass*, on platinum, it melts in larger quantity than with soda, but the fused mass is not so liquid ; the glass when cold is black with no admixture of red.—With a small portion of *soda*, on charcoal, in the reducing flame, no fusion takes place ; but reduction is very easily effected ; we obtain a grey metallic magnetic powder.

----

278. We now come to describe an order of metals, some of the combinations of which with oxygen are possessed of properties which entitle them to the appellation of *acids*.—We shall give the characters by which these curious substances are respectively distinguished ; but with regard to the bodies which bear to them the relation of SALTS, that is to say, the *arseniates, chromates, molybdates, tungstates,* &c., it is merely necessary to say, that when the habitudes of the *oxides*, or *acids*, are known, those of their *salts*, or

alloys, will be easily understood. Nevertheless, we have described the phenomena exhibited by several of these compounds.

TELLURIUM.

279. This is an extremely rare metal. It is found only in the native state; but it is always alloyed with other metals.

280. *Tellurium, in the metallic state, alone*, in the matrass, first gives off a vapour, and then a sublimate of grey metallic tellurium. In the open tube, it produces copious fumes which adhere to the glass as a white powder, but can be melted into clear colourless drops. On charcoal, it burns with a vivid blue light, greenish on the edges; flies off at a gentle heat in greyish white fumes, which condense into a white oxide; leaves a mark on the support which tinges the reducing flame, when directed upon it, of a fine deep green colour. In the *matrass*, it melts into a straw-coloured striated mass.—It forms a deep purple-coloured solution with muriatic acid.

281. *Oxide of Tellurium, alone*, on platinum foil, melts and fumes. On charcoal, when gently heated, it becomes first yellow, next light red, and afterwards black. It then melts, is absorbed by the charcoal, and finally is reduced. The reduction is attended by effervescence, a slight detonation, and a greenish flame. Oxide of tellurium frequently exhales the odour of putrid horse-radish, which is an indication of the presence of selenium.

282. With *borax* or *mic. salt*, on the platinum wire, it gives a clear colourless glass, which becomes grey and opaque on charcoal.—With *soda*, on the platinum wire, it forms a colourless glass, which turns white on cooling. On charcoal, by the reducing process, it is converted into metallic tellurium.

283. *Alloys of Tellurium*, in which the oxide of this metal appears to act the part of an *acid*, roasted in the open tube, behave like metallic tellurium.

The vapour produced has a pungent odour peculiar
to itself; if it smells of horse-radish, it is a proof
that selenium is present.

ARSENIC.

284. This is a metal of very frequent occurrence.
It is found in the native state, also alloyed with several metals, particularly silver, cobalt, copper, and
antimony; and combined with lime, sulphur, and
oxygen. In some of its combinations, it acts, in conjunction with oxygen, the part of an *acid*.

285. *Arsenic, in the metallic state,* is detected by its
odour when heated, which resembles the smell of
garlic. It is among the most combustible of the
metals; burns with a blue flame at a low heat, and
sublimes in the state of arsenious acid.

286. *Arsenious acid,* or white arsenic, or white
oxide of arsenic, which all mean the same thing,—
when heated in large pieces on ignited charcoal,
exhales no smell: hence it appears that it is not
arsenious acid, but metallic arsenic in a volatile
state, that emits the odour of garlic. But if the
white oxide is reduced, by being mixed and heated
with powdered charcoal, *then* it does exhale the
odour of garlic.

287. By a proper quantity of oxide of arsenic, the
fluxes are rendered yellow without opacity; and by
a long continued heat, the volatile additament is expelled. Iron and copper precipitate arsenic in the
metallic form.

288. If arsenic is held in *solution,* it may be discovered by dipping into the solution a piece of pure
and well-burned charcoal, which is afterwards to be
dried and ignited.

289. *Alloys of Arsenic.* The roasting of these
should be begun in the open tube; to the sides of
which the arsenious acid attaches itself in the form
of a white crystalline sublimate; sulphurous acid, if
any be present, then rises by itself from the tube,

and is easily perceived. The roasting may be completed on charcoal, using the interior and exterior flame alternately. If arsenic be present in an assay in large proportion, its characteristic odour diffuses itself to a considerable distance. If the proportion be very small, the odour may probably be only developed by the reducing process. See paragraph 131 for several useful remarks on this subject.

CHROMIUM.

290. This has never been found in the metallic form, either alone, or alloyed with other metals. It is generally found combined with oxygen, and forming sometimes an acid, sometimes an oxide. It is the colouring matter of the emerald; but its principal ores are the chromate of lead and the chromate of iron.

291. *Oxide of Chromium.*—*Alone*, on charcoal, or platinum, its *green oxide*, (the protoxide,) the form in which it most commonly occurs, and to which *columbic acid* is reduced by heating in the common air, is infusible and unalterable.

292. With *borax*, in the exterior flame, it melts with difficulty, and forms a bright yellowish or yellow-red glass; in the interior flame, this becomes darker and greener, or blueish green, retaining a green tinge on cooling.—With *mic. salt*, it melts either in the external or internal flame, and the glass assumes in both cases a blueish green colour, the intensity of which varies with the proportion of the assay. A deep tint is produced by a very small portion of the oxide. The glass formed in the interior flame, at the instant of its removal, has a violet hue.—With *soda*, on the platinum wire, in the external flame, it produces a dark orange glass, which turns yellow and opaque on cooling; in the reducing flame, it gives an opaque glass, which, on cooling, is green. It does not yield metallic chromium by the reducing pro-

L

cess, although the assay is absorbed by the charcoal.

MOLYBDENUM.

293. This is a very rare metal. It has never been found pure, but occurs combined with sulphur, also, forming with oxygen an acid which enters into the composition of a mineral called molybdate of lead.

294. The metallic globules of molybdenum are extremely infusible. But they are converted by heat into a white oxide which rises in brilliant needle-formed flowers, like those of antimony.

295. *Molybdic acid, alone,* on charcoal, melts with ebullition, is absorbed, and by continuing the blast, partly reduced. By washing (164) a grey metallic powder may be obtained. On platinum, it melts and emits white fumes; in the interior flame it is reduced to molybdous acid, which is of a beautiful blue colour; in the exterior flame it is again oxidated, and becomes white; a violent heat makes it brown. In the open tube, it melts and produces fumes, which partly condense as a white powder on the sides of the tube, and partly as brilliant pale yellow crystals on the surface of the melted assay.

296. With *borax*, on charcoal, in the interior flame, it fuses, and a quantity of blackish scales of oxide of molybdenum are precipitated from the clear glass, which is left colourless when the quantity of molybdenum is small, and blackish when the proportion is larger.—With *mic. salt*, on the platinum wire, or charcoal, in the exterior flame, a small proportion of the acid gives an elegant green glass, which by gradual additions of the acid, passes through yellow-green to reddish, brownish, and hyacinth-brown with a slight tinge of green. In the interior flame, the colour passes from yellow-green, through yellow-brown and brown-red, to black; and if the proportion of acid be large, it acquires a metallic lustre, resembling that of the sulphuret, which lustre is some-

times retained after the glass has cooled.—With soda, it dissolves with violent effervescence; the glass is red and transparent; on cooling it becomes paler in colour and opaque, and acquires an hepatic smell. Molybdic acid yields molybdenum, in the state of a steel grey metallic powder, by the reducing process. See paragraphs 162–6.

TUNGSTENUM.

297. This is by no means a plentiful metal. It occurs combined with oxygen forming an *acid*, which with lime constitutes native tungstate of lime, and with iron, native tungstate of iron.

298. *Tungstic acid, alone*, on platinum, or charcoal, becomes at first brownish yellow, is then reduced to a brown oxide, and lastly, becomes black, without melting or smoking.

299. With *borax*, on the platinum wire, it melts with ease in the exterior flame, and forms a colourless glass, which is not made opaque by flaming. On charcoal, with borax, in the internal flame, and in small proportions, it forms a colourless glass, which, by increasing the proportion of the acid, becomes dirty grey, and then reddish. By long exposure to the external flame, it is rendered transparent, but, as it cools, it becomes muddy, whitish, and changeable into red when seen by day-light. If *tin* is added, the glass forms an enamel on cooling.—With *mic. salt*, in the interior flame, it gives a fine pure blue glass, more beautiful than that formed with cobalt; in the exterior flame this colour disappears; but returns again in the interior flame. If *iron* be present, the assay, instead of becoming blue, assumes a blood-red colour; in which case, the addition of a little tin, destroys the effect of the iron, and causes the glass to have either a green or a blue colour. In the oxidating flame, it gives a yellowish glass.—With *soda*, on the platinum wire, it forms a yellow semi-transparent glass, which on cooling crystallizes and

turns opaque and white, or yellowish. With soda,
on charcoal, by the reducing process (162), we obtain metallic tungsten in the form of powder possessing a steel grey colour, and a partial metallic brilliancy.

COLUMBIUM.

300. This is an extremely rare metal. It occurs
combined with oxygen forming an oxide or acid.
Its ores are only two in number. In one of these it
is combined with iron, in the other with the rare
earth yttria.

301. *Oxide of Columbium*, or Columbic Acid, on
platinum, undergoes no change.

302. With *borax*, it forms a colourless transparent
glass, which flaming renders opaque; if too much
oxide be present, the glass forms a white enamel on
cooling.—With *mic. salt*, it forms a glass which is
perfectly transparent.—With *soda*, it effervesces and
combines, but neither melts nor is reduced.—With
solution of cobalt, it does not produce a blue colour.

SELENIUM.

303. This name was given by Berzelius to a substance which he discovered in the iron pyrites of Fahlun. Its properties are so peculiar that chemists are
not decided whether it should be classed with metals or combustibles. It is extremely scarce. It
combines with a few metals forming *seleniurets*.

304. When heated, it softens; at 212° it is semi-
liquid; and at a temperature a few degrees higher,
it melts completely. It cools very slowly, and is
meanwhile exceedingly soft and ductile. Before the
blowpipe, it is easily volatilized, and gives out a
smell of decayed horse-radish, which is of so strong
a nature that the odour produced by the conversion
of the fifteenth part of a grain into the gaseous oxide,
is sufficient to scent a large apartment.

305. All the metallic compounds into which sele-

nium enters are easily distinguished by the horse-
radish odour they emit, when heated in the oxidating flame; for this small is so strong and disagreeable that the smallest portions of selenium may be
detected by it. When seleniurets are roasted in the
open tube, selenium in a state of purity often sublimes in the form of a red powder. If sulphur be
present, it escapes in the state of sulphurous acid
gas, without affecting the sublimation of the selenium. If tellurium be present, it sublimes previously to selenium, and is known by its peculiar characters.

306. The combination of the following seven metals with
oxygen, we have described as being *not much known*. What
we have said of the oxides, we might just as well have said of
the metals; for not one of them has ever been obtained in
any considerable quantity, or had its properties investigated
with any great minuteness, or been usefully applied to any
purpose of importance. Still it is necessary to know how
to detect them when they occur in minerals.

OSMIUM.

307. Exceedingly rare. Occurs alloying native
platinum, from which it is obtained in the metallic
state with great difficulty.

308. *Alone,* in a gentle heat, it is converted into
an oxide, which immediately volatilizes, giving a
peculiar pungent odour, something like that of chlorine.

RHODIUM.

309. A scarce metal, which occurs alloying platinum.

310. *Alone*, it is infusible.

311. With *fluxes* it does not act.

312. Like platinum, it is best examined by cupellation, which clears it from foreign bodies, and yields
a grey, porous, infusible, metallic mass.

IRIDIUM.

313. The description of rhodium applies in every respect to iridium.

URANIUM.

314. This is a very scarce metal. It has never been found pure. Its ores are few. It occurs in one, in the state of an oxide. In another, in that of a phosphate. It is also found mineralized by sulphur.

315. It is extremely difficult of fusion, and can only be procured in the metallic state by a laborious and complicated process. It never has been reduced to the metallic state but in very small quantities.

316. *Oxide of Uranium.*—*Alone*, on charcoal, the yellow oxide becomes greyish black, but does not fuse.

317. With *borax*, in the interior flame, it forms a clear, colourless, or faintly greenish glass, containing black particles, which appear to be the metal in its lowest state of oxidation. In the exterior flame, this black matter, if the quantity be not great, is dissolved, and the glass becomes first bright yellowish green, and after farther oxidation, yellowish brown. If brought again into the reducing flame, the colour gradually changes to green, and the black matter is again precipitated, but no farther reduction takes place.—With *mic. salt*, on the platinum wire, in the external flame, it gives a clear glass, yellow while hot, pale greenish yellow when cold. On charcoal, in the reducing flame, it gives a fine green glass, the colour of which, on cooling, is increased in beauty.— With *soda*, on charcoal, it acquires a yellowish brown colour, but does not dissolve nor reduce.

TITANIUM.

318. A rare metal. It occurs commonly in the

state of an oxide, often combined with iron, lime, or silica.

319. It is exceedingly difficult of fusion and reduction.

320. *Oxide of Titanium, alone,* on charcoal, when ignited, turns dark-brown; in a platinum spoon, it turns yellow. It does not melt.

321. With *borax,* on the platinum wire, it melts easily, and gives a colourless glass, which flaming renders milk white; in the reducing flame, the glass assumes a dull amethystine hue. On charcoal, in the internal flame, it melts in large quantity; the glass produced is dull yellow while hot, deep blue when cold.—With *mic. salt,* in' the oxidating flame, it gives a transparent colourless glass; in the interior flame, produces a glass which is yellow while hot, red while cooling, very fine blueish violet when cold.—Oxide of titanium, with *mic. salt,* and *iron,* in the reducing flame, yields a glass of a red colour.—With *soda,* it effervesces and melts into a dull yellow glass, which crystallizes when the igniting blast is suspended, and evolves a prodigious degree of heat. Metallic titanium is not attainable by the reducing process.—With *solution of cobalt,* it becomes greyish black.

CERIUM.

322. This is an extremely rare metal. In the cerite and allanite, it occurs in the state of an oxide. In the yttrocerite, and a few other rare minerals, it is found combined with fluoric acid.

323. It is so very difficult of fusion and reduction, that it can scarcely be said to have been yet obtained in the form of a metal.

324. *Oxide of Cerium, alone,* on charcoal or platinum, becomes brown-red, but does not fuse.

325. With *borax,* a small proportion of the assay heated in the internal flame, produces a glass, faint yellow-green while warm, colourless when cold; a

large proportion of the assay, gives a glass which, on cooling, becomes crystalline and enamel white. In the oxidating flame, it forms a glass of a beautiful red, or deep orange colour, which fades to a pale yellow when cold.—With *mic. salt*, it produces a glass, fine red while hot, colourless and perfectly transparent when cold.—With *soda*, does not fuse nor reduce; the flux is absorbed, leaving a grey-white powder on the charcoal.

WODANIUM.

326. A scarce metal, said to have been found in a mineral called wodan pyrites; in which it is combined with sulphur, arsenic, iron and nickel. It has not been tried by the blowpipe. Some chemists consider the existence of this metal to be a matter of doubt. It rests on the credit of Lampedius.

POTASSIUM :—SODIUM :—LITHIUM,

327. These three metals are only capable of being obtained in the metallic state by the powerful agency of the voltaic battery. They are, therefore, merely to be considered here as the bases of the three *fixed alkalies*, which are binary compounds of the above substances with oxygen.

POTASS is composed of *Oxygen* and *Potassium*.

SODA   ~   ~   ~   ~   *Sodium*.

LITHIA   ~   ~   ~   ~   *Lithium*.

These alkalies form the bases of several saline minerals, and enter into the composition of some of the earthy minerals. Lithia is of very rare occurrence.

328. Before the blowpipe, the alkalies do not present any phenomena which can be said to characterise them. They are best distinguished by means of tests. A fragment of a mineral containing an alkali, if heated to redness, and placed on brazil-wood paper, or litmus paper reddened by an acid, forms a blue stain round the spot where it lay.

POTASS.

329. This is not a plentiful substance in the mineral kingdom; it occurs in about twenty earthy minerals, among which are felspar and mica. It is also found in combination with the carbonic and nitric acids; but does not enter into the composition of the metalliferous substances. Nitro-muriate of platinum produces in a solution of the salts of potass an orange-coloured precipitate; but has no such effect with solutions of the other alkalies.

330. *Crystallized Carbonate of Potass*, before the blowpipe, first becomes opaque, and decrepitates long and violently, then melts into a globule which retains its form on platinum, but on charcoal expands, and is absorbed with a crackling noise.

SODA.

331. This is a constituent of about fifteen earthy minerals in proportions varying from 1 to 35 per cent. It also occurs combined with the sulphuric, carbonic, boracic, and muriatic acids; but is not found in metalliferous substances. It is much more abundant than potass.

LITHIA.

332. This is a very rare substance. It occurs in the petalite, spodumene, and a variety of the lepidolite. Before the blowpipe, it is distinguished from the preceding alkalies by the following circumstance. When heated to redness, on platinum foil, it corrodes that metal, and forms a dull yellow trace round the spot where it lay.

AMMONIA.

333. This, which is often called the volatile alkali, is composed of hydrogen and nitrogen, and is placed here merely on account of its possessing alka-

line properties. It only occurs in the mineral kingdom combined with sulphuric and muriatic acid.
We detect it by the odour which is produced when
a substance which contains it is mixed with soda.
Before the blowpipe, it is entirely dissipated.

---

234. We come now to the consideration of the last order
of metals, or rather to the consideration of THE EARTHS,
which are the substances formed by the combination of the
different remaining metals with oxygen. For as these metals are only separable by means of the most powerful galvanic agency, and scarcely indeed with that, it is only with
their oxides that we have any thing to do.—The Earths are
ten in number. In the following list, they are placed in the
order of their abundance in the mineral kingdom; that
which occurs most plentifully being placed at the top, and
being followed by the others in regular order:—

Earths.	Metals of which they are Oxides.
Silica	Silicium
Alumina	Aluminum
Lime	Calcium
Magnesia	Magnesium
Zirconia	Zirconium
Glucina	Glucinum
Yttria	Yttrium
Barytes	Barium
Strontia	Strontium
Thorina	Thorinum

The Earths, as is indicated by the name, form the principal part of Earthy Minerals; sometimes occuring
nearly pure, but more frequently in combination either
with other earths, or alkalies, or acids. The metallic
matter in earthy minerals, and to which the colour they
possess is generally owing, is in most cases an acciden-
al admixture.

Of the earths, Silica, Alumina, Lime, and Magnesia, are by far the most common, both in simple minerals and in aggregated masses or rocks; silica being
the most and magnesia the least abundant of the four.—
Siliceous and Aluminous earths are most frequently
combined with other earths, and with the alkalies potass
and soda; but the proportion of the alkaline ingredient
is in some minerals so small, that we know not whether

to esteem it an essential or accidental part.—Lime is generally combined with sulphuric or carbonic acid, and more rarely with the fluoric or phosphoric ; it is also frequently combined with magnesia, but much more sparingly with the other earths.—Magnesia is sometimes combined with silica and alumina, but more generally with lime.

The other six earths occur in much smaller quantities than the preceding. Barytes and Strontia, are rare ; they occur in veins. Zirconia exists in the Hyacinth and Zircon. Glucina occurs in the Emerald, Beryl, and a few other minerals. Yttria and Thorina are found in a few minerals of great rarity.

SILICA.

335. This is esteemed to be the most abundant substance in nature. It is the chief ingredient in the greatest masses of rocks and soils. It occurs in the hardest gems, and the softest clays. By analyses, it has been found, in various proportions, in about two-thirds of the whole number of earthy minerals whose composition is known. It does not occur combined with acids, although it is sometimes contained in acidiferous earthy substances. It appears, indeed, in many cases, to behave as an acid itself.

336. *Alone*, on platinum, or charcoal, it is infusible.

337. With *borax*, it melts slowly without effervescence ; gives a clear glass, which is difficultly fusible and not rendered opaque by flaming.—With *soda*, it effervesces briskly, and fuses into a clear glass.—With *mic. salt*, it dissolves partially and very slowly, without effervescence ; the residue being rendered semi-transparent ; the glass which is formed is permanently transparent.—With a small quantity of *solution of cobalt*, it becomes pale blue when perfectly fused ; with more solution of cobalt, black ; thus, silica is distinguished from alumina. See paragraphs 170 and 343.

338. *Silicates.* To the combinations of silica with

other earths, Berzelius has applied the term *silicates*,
stating that these combinations are in fact salts, in
which silica plays the part of an acid towards the
bodies with which it is combined.

339. With *mic. salt*, a silicate is decomposed;
silica is disengaged; the base combines with the
phosphoric acid of the flux. If the assay is large in
proportion to the flux, it often swells up and absorbs
the fused mass. If the flux is large in proportion to
the assay, a globule is formed, holding a tumefied
translucent mass of silica in suspension. But this is
only clearly seen when the glass is examined while
hot; for most silicates give glasses, which, though
transparent while in fusion, become nearly opaque
when cold. If the assay contains but a small quantity of silica, it generally dissolves in the flux entirely.

340. With *soda*, the silicates develope different
phenomena; acting according to the relation which
exists between the quantity of oxygen in the silica,
and the quantity of oxygen in the base which is
combined with the silica. This, at least, is the opinion of Berzelius, to whose works we refer for an
enunciation of his theoretical views upon this difficult subject. We have no room for discussion here.
The following facts are therefore given without any
attempt at explanation :—1. Every earthy or stony
substance which effervesces with soda, and melts
into a permanently transparent glass, is silica, or a
silicate. 2. Decomposition and the formation of a
glass sometimes partially take place; but the unfused part of the assay absorbs the glass into its pores.
3. A silicate will often form a clear glass with a
small portion of soda; an opaque one with more
soda; and become infusible with a quantity of soda
still larger.

ALUMINA.

341. This earth is the basis of all the clays or

argillaceous substances, and is, therefore, one of the
most plentiful of mineral constituents. Indeed, it is
the most abundant of all the earths, except silica.
One of the simplest characters by which aluminous
minerals are distinguished, is, that when breathed
upon, they give out a peculiar earthy odour. · Alumina occurs in metallic minerals in small quantities.

342. *Alone*, heated on platinum or charcoal, it is
infusible, but contracts and becomes hard.

343. With *borax*, it slowly melts into a transparent
glass, which is not rendered opaque, either by cooling or flaming. If pounded alumina be added in
large quantity to the glass while hot, it mixes with
it and makes it opaque, but does not fuse.—With
soda, it swells, and forms an infusible compound.—
With *mic. salt*, it fuses into a clear glass, which saturation does not render opaque.—With *solution of
cobalt*, after having been dried and strongly ignited
for some time, it gives a beautiful bright blue colour,
which becomes deeper, without losing its beauty, by
an additional quantity of cobalt. This is the most
striking character of alumina, the presence of which
in compound minerals may in this manner be easily
detected; unless they contain also a great proportion of metallic substances, or a large quantity of
magnesia. The blue colour is only distinctly seen
by day-light, and when the assay is perfectly cold.
See paragraphs 170 and 337.

344. Bergman says, that Alumina effervesces with
soda, " a little ;" with *borax*, " remarkably ;" with
mic. salt, " still more violently." Neither Gahn
nor Berzelius say anything about the effervescence
of alumina with the fluxes. Bergman used pure
clay from earth of alum in his experiments.

LIME.

345. This substance is exceedingly abundant in
the mineral kingdom, not in a pure state, but generally in combination with acids, such as the carbo-

nic, sulphuric, fluoric, and several others, with which
it forms acidiferous earthy minerals. Carbonate of
lime or calcareous spar, sulphate of lime or gypsum,
and fluate of lime or fluor-spar, are minerals of this
kind of frequent occurrence.

846. *Alone*, on charcoal, or platinum, it neither
melts nor suffers any change.

347. *Carbonate of Lime*, is easily rendered caustic,
by the volatilization of its acid; at the same time
emitting a brilliant light. It evolves heat and falls
to powder on being moistened. This may be easily
tried by placing a particle of lime, just cooled after
burning, on the back of the hand, and adding a drop
of water. The *hydrate* thus found is infusible.

348. With *borax*, lime melts into a clear glass,
which flaming renders opaque; the carbonate fuses
with effervescence; a larger quantity of lime (either
the carbonate or the pure) gives a transparent glass
which on cooling crystallizes; the glass of lime is
never so perfectly milky as those formed with barytes and strontia.—With *soda*, it is infusible, but
the carbonate effervesces and is divided into particles.
—With *mic. salt*, lime fuses in large quantity; forms
a glass which remains transparent when cold; the
carbonate does the same, but effervesces considerably
during fusion: if the glass is supersaturated with
lime, and is exposed to a long-continued blast, it
then turns milk white on cooling.—With *solution of
cobalt*, it gives a dark grey infusible mass.

849. " It is observable," says Bergman, " that a very
small piece of calcareous earth is easily dissolved in borax or
microcosmic salt, yielding a spherule altogether pellucid;
but if more earth be gradually added, the flux, at length saturated, retains the dissolved matter, indeed, while in perfect
fusion; but, on removing the flame, the part which was taken
up by means of the heat alone, separates; hence, clouds
arise at first, and the whole globule becomes opaque, but
recovers its transparency again by fusion. This is entirely
correspondent to what happens in the humid way. For in
water, saturated with nitre, or Glauber's salt, upon cooling,

is obliged to deposit that part which it had taken up in virtue of its warmth. If the fused pellucid globule, which would grow opaque upon cooling, be quickly plunged into melted tallow, water, or other substance, hot, (for cold generally cracks it,) so as to grow suddenly hard, it retains its transparency; the particles being as it were fixed in that state which is necessary to transparency. This is a phenomenon highly worthy of observation, and which cannot be seen in the crucible.

350. *Sulphate of Lime*, or Gypsum, is easily reduced to sulphuret, and possesses, besides, the property of combining with fluor-spar, forming a clear glass, in a moderate heat. See paragraph 173.

MAGNESIA.

351. The substances in which this earth generally occurs are serpentine, basalt, and magnesian limestone. It is also found, in rather small quantities, in about thirty earthy minerals, and in a few minerals of the metalliferous class. It also occurs combined with the carbonic, sulphuric, and boracic acids.

352. *Alone*, on platinum, or charcoal, it undergoes no change; but produces an intense brightness in the blowpipe flame.

353. With *borax*, it behaves like lime.—With *soda*, it has no action.—With *mic. salt*, it fuses readily; a small quantity forms a clear glass which becomes opaque by flaming; a large quantity gives a glass which on cooling spontaneously becomes milk white.

354. With *solution of cobalt*, when strongly ignited, a faint reddish colour like flesh is produced. The tint, however, is only distinguishable when the assay is perfectly cold, and by day-light. Magnesia may by this process be detected in compound bodies, in which it occurs in very small proportions, provided they do not contain much metallic matter, or a proportion of alumina exceeding that of the magnesia. Some inference as to the quantity of magnesia

contained in a mineral may generally be drawn from the intensity of the colour produced by the assay. See paragraph 170.

355. *Carbonate of Magnesia, alone,* loses its acid. —With *borax,* it effervesces slightly, and behaves like pure magnesia.—With *mic. salt,* the effervescence is more violent.—With *soda,* it effervesces a little, and is somewhat diminished.

ZIRCONIA.

356. Occurs very sparingly; found only in a few minerals.

357. *Alone,* on platinum, or charcoal, it is infusible; but is remarkable for emitting a light of intense brilliancy.

358. With *borax,* it behaves like glucina.—With *soda,* the same.—With *mic. salt,* the same; but dissolves more sparingly, and the glass more easily becomes opaque.

GLUCINA.

359. This is found only in small quantities, and in a very few minerals, namely, euclase, beryl, emerald, gadolinite, and topazolite.

360. *Alone,* on platinum, or charcoal, it experiences no change.

361. With *borax,* it dissolves in large proportion; forms a clear glass, which flaming renders milk white; if supersaturated with the assay, the glass spontaneously becomes milk white on cooling.— With *soda,* no action ensues.—With *mic. salt,* as with borax.—With *solution of cobalt,* it forms a dark grey mass.

YTTRIA.

362. Of extremely rare occurrence.

363. Its habitudes before the blowpipe resemble those of glucina.

BARYTES.

364. This is rather a scarce substance. It occurs combined either with the sulphuric or carbonic acid, forming compounds which may be readily distinguished from most other earthy minerals, by their great specific gravity.

365. *Alone,* on charcoal, or platinum, it is infusible; but when containing water, it intumesces and melts, and if on charcoal, penetrates it, and forms a solid crust.

366. *Carbonate of Barytes,* on platinum, very easily melts into a clear glass, which becomes a white enamel on cooling; on charcoal, it melts easily, effervesces strongly, sputters up, becomes caustic, and sinks into the support.

367. With *borax,* a small quantity of either of the above, effervesces briskly, and fuses into a clear glass which becomes opaque by flaming; a larger portion of the assay produces a glass which turns enamel-white on cooling, without flaming.—With *mic. salt,* the same phenomena take place as with borax, but foams and intumesces; ends in a clear glass, or a white enamel, according to the proportion borne by the assay to the flux.—With *soda,* on charcoal, it melts and is absorbed.—With *solution of cobalt,* it gives a reddish brown globule; the colour of which flies on cooling.

368. *Sulphate of Barytes* is converted, in the interior flame, into a sulphuret, and is absorbed by the charcoal, with effervescence, which is continued as long as it is exposed to the action of the blowpipe.

STRONTIA.

369. This is a scarce earth. It occurs in the two states of carbonate and sulphate.

370. *Alone,* on platinum, or charcoal, either pure

or in the state of a *hydrate*, (that is, containing water,) it behaves like barytes.

371. *Carbonate of Strontia*, held in small thin plates, with platina forceps, in the reducing flame, melts on the surface; the carbonic acid is driven off, and the assay assumes a ramified cauliflower appearance, possessing a dazzling brightness. In an intense heat, and on the side of the plate farthest from the lamp, a red flame is seen, sometimes edged with green, and scarcely perceptible but by the light of a lamp. The ramified portion is alkaline.

372. With *borax* and *mic. salt*, it behaves like barytes.—With *soda*, caustic strontia remains unchanged, but the carbonate in small quantity melts into a clear glass which becomes white on cooling, or if strongly heated, bubbles and is absorbed; a large quantity of the carbonate only fuses partially. —With *solution of cobalt*, it turns black, but does not melt.

373. *Sulphate of Strontia* is reduced in the interior flame to a sulphuret. Dissolve this in a drop of muriatic acid, add a drop of alcohol, and dip a small bit of stick in the solution: it will burn with a fine red flame.

THORINA.

374. This earth has only been found in a few rare minerals. It occurs in the fluates of cerium and yttria, and sometimes in the gadolinite. We are but little acquainted with its properties.

375. *Alone*, it is infusible.

376. With *borax* or *mic. salt*, it fuses into a transparent glass.

ACIDS.

377. The foregoing descriptions of the characteristic phenomena presented by metallic substances before the blowpipe, will enable any one readily to ascertain the nature of the BASE of any *saline compound* which he may have to

examine. It is necessary to show now, by what means an Acid may be distinguished.

378. The acids which enter into the composition of minerals are about thirteen in number. Of these, four are metallic acids, and have been already treated of: namely, the tungstic, chromic, molybdic, and arsenic acids. Only nine, therefore, remain to be described; these are the following:—

Acids.	Elements of which they are composed.		
Carbonic	Carbon	and	Oxygen
Phosphoric	Phosphorus		Oxygen
Fluoric	Fluorine		Hydrogen
Sulphuric	Sulphur		Oxygen
Muriatic	Chlorine		Hydrogen
Nitric	Nitrogen		Oxygen
Boracic	Boron		Oxygen

Succinic — This is the acid obtained from *Amber*. It is a triple compound containing Carbon, Oxygen, and Hydrogen.

Mellitic — This is the acid obtained from *Honeystone*. Its nature is unknown.

The carbonic, phosphoric, fluoric, sulphuric, muriatic, nitric, and boracic acids occur combined with *earths*.—The carbonic, phosphoric, sulphuric, and muriatic acids are found mineralizing certain *metals*.—The nitric and carbonic acids are found united with *potass*.—The carbonic, sulphuric, muriatic, and boracic acids occur in combination with *soda*.

379. *Carbonic Acid.*—The blowpipe affords no method of detecting this so good as the common one of a drop of nitric or muriatic acid, by means of which effervescence is produced.

380. *Phosphoric Acid.*—This occurs in combination with various earths and metals, forming *phosphates.* It is found also, in very small proportions in wavellite, lazulite, and a few other minerals. Now, as phosphoric acid has the property of being precipitated with the earthy bases, in experiments

performed in the wet way; it consequently frequently eludes the researches of the chemist when he operates in that manner. Hence a test to detect it in the dry way becomes the more desirable. The following will be found an effectual method:—Melt the assay with boracic acid, on charcoal, and plunge into the globule, while it is hot, a bit of steel wire, a little longer than the diameter of the globule. Then heat the whole in a good reducing flame. If the assay contain less than a twentieth part of its bulk of phosphoric acid it will not be rendered evident; the wire will burn at the projecting ends, but preserve elsewhere its metallic brilliancy. But if a greater proportion of phosphoric acid be contained in the assay, it is decomposed; and there is simultaneously formed borate of iron and phosphuret of iron. The latter melts in a strong heat, and the assay resumes its globular figure. When it is cold, the globule may be wrapped in a piece of paper, and struck with a hammer: when there will be found among the fragments a metallic globule more or less brittle, magnetic, and with a fracture having a steel colour. This is the phosphuret of iron. It is necessary, however, to observe, that this experiment does not succeed if any other substance be present in the assay, capable of being reduced and fused with iron; such as the sulphuric or arsenic acid.

381. *Fluoric Acid.*—This is characterised by its peculiar odour and by its properties of corroding glass, and of changing the red colour of brazil-wood test paper to yellow. If a small piece of a fluate be moistened in a watch glass with muriatic acid, the mixture after a few seconds, produces its peculiar effect when rubbed on brazil-wood paper.—If a mineral containing a large quantity of fluoric acid, be mixed with fused mic. salt, and heated in the open tube, fluoric acid gas is formed. This may be known by its peculiar odour, by its corroding the interior surface of the glass tube, and by the effect

produced on a piece of b●zil-wood paper inser●
into the upper part of the tube. This plan is best
adapted for the minerals of which fluoric acid is an
essential part, such as fluor-spar, topas, cryolite, &c.
—Mica, and other minerals, containing a small portion of fluoric acid, as an accidental ingredient, are
best tried in a small matrass, when the usual
effects are developed, though in an inferior degree.

382. *Sulphuric Acid.*—A globule is formed by
melting together silica and ●oda, and the mineral to
be tried is placed on this. Or the assay is mixed
with the soda, previous to its fusion with the silica.
The first method produces the most accurate result;
the second is the easiest process. In either case, if
sulphuric acid is present, and if the assay fuses
when exposed to the flame, the sulphuric acid is decomposed, and sulphuret of soda is produced. This
is known by the dark orange or reddish brown tinge
acquired by the glass, either while hot or when
cold.

383. *Muriatic Acid.*—To detect this, oxide of
copper and mic. salt are melted into a dark green
globule. The assay is added to this, and the whole
is exposed to the blast: if muriatic acid be present,
the globule becomes surrounded by an elegant blueish purple flame, which continues as long as any
muriatic acid remains in the assay.—*Iodates* treated
in the same way as *muriates* produce a superb deep
green flame.

384. *Nitric Acid.*—The fusible nitrates detonate
with charcoal. The infusible nitrates, when gradually heated, till they are red hot, in a small matrass,
give out the orange coloured vapour of nitrous acid.

385. *Boracic Acid.*—No flux or re-agent has yet
been discovered capable of detecting in blowpipe
experiments minute quantities of this substance,
which is the more to be regretted, as it frequently
occurs in minerals in minute proportions.

HYDRATES.

384. Nearly all minerals, when heated in the matrass, give off more or less water, the slightest trace of which condenses in the upper part of the apparatus. In most cases, water exists in minerals as an accidental ingredient, especially in those of a porous texture. But in a few substances it is an essential constituent part, as is evinced by the difference existing between the forms of the primary crystals of hydrous, and those of anhydrous, sulphate of lime.

A

SYSTEMATIC ARRANGEMENT
of
SIMPLE MINERALS,

FOUNDED UPON

THEIR MOST IMPORTANT CHARACTERS,
PHYSICAL AND CHEMICAL;

AND ADAPTED TO

AID THE STUDENT IN HIS PROGRESS IN
MINERALOGY,

BY FACILITATING THE DISCOVERY OF THE

Names of Species.

---

887. As the nature of the following System has been fully explained in the Introduction, it is unnecessary to trouble the reader with any remarks here. We shall proceed at once, therefore, to lay before him the details of the plan which is there unfolded.

388. In the first place, we shall give the Essential Characters of all the classes collectively, and then proceed to the description of each particular class.

---

389. CLASS 1. COMBUSTIBLE MINERALS.
Essential Characters :—

The substances of this class are of *low specific gravity*, scarcely exceeding 2. 0. (water being = 1. 0.) when pure; none are above 3. 5. (except the diamond); and some are supernatant.

They are all *soft ;* some liquid ; the hardest yield with ease to the knife,—except the diamond.

Some of them are eminently *combustible,* burning away at, or below, a red heat.

The rest are combustible with greater or less ease, and more or less completely, by the action of the blowpipe.

The diamond, which differs so remarkably in its specific gravity and hardness from the other minerals of this class, is slowly combustible at a heat rather below the melting point of silver.

———

390. CLASS 2. METALLIC MINERALS.

Essential Characters :—

Specific Gravity exceeding 5. 0....................................

Possess a Metallic Lustre when scraped................... } *Minerals which possess either of these characters are metallic.*

Or,

Specific gravity less than 5. 0. but more than 2. 5., and destitute of metallic lustre, but

a. Reducible to the metallic state by the blowpipe, or

b. Rendered Magnetic by the blowpipe, or

c. Volatilized, wholly or in part, by the blowpipe, producing a vapour, or

d. Communicate a colour to borax by the blowpipe.

Some of the substances in this class, like the non-metallic minerals of class 1, are inflammable. No confusion, however, can arise hence; because all such substances greatly exceed in specific gravity the heaviest of the combustible minerals.—The specific gravity of the minerals constituting class 1, very rarely, indeed, rises so high as 2·5., while the specific gravity of those of class 2, as seldom descends so low.

———

391. Class 3. EARTHY MINERALS.

Essential Characters :—

Insoluble in water ⎱ *These characters distinguish earthy*
Insapid................ ⎰ *minerals from saline.*

Incombustible at a ⎱ *This character distinguishes earthy*
white heat........... ⎰ *minerals from combustible.*

Specific gravity less than 5·0.....⎫
Destitute of true metallic lustre* ⎪ *These characters*
Neither Reducible to the metallic ⎬ *taken together*
   state, nor Volatilizable at a high ⎪ *distinguish*
   temperature, before the Blow- ⎪ *earthy from*
   pipe.......................................⎭ *metallic minerals*

———

392. Class 4. SALINE MINERALS.

Essential Characters :—

Soluble in water........⎱ *These characters are sufficient to*
Taste sapid................⎰ *distinguish saline minerals from all others.*

——————————

* A few earthy minerals have a metallic appearance, which arises from their possessing a false or pseudo-metallic lustre. But the illusion is destroyed by scraping the surface of such minerals with a knife, upon which the lustre vanishes; whereas, minerals possessing the true metallic lustre, become brighter when they are scraped.

N

393. All the above *Classes* of Minerals are divided into ORDERS, most of these *Orders* into GENERA, and the *Genera* again into FAMILIES; every successive division being founded on characters that are peculiar to the bodies which constitute it; wherefore, an account of these characters is placed at the head of every division. By this means, the enquirer is directed straight onwards to the object he is in search of, like the traveler who meets at every corner with signs to direct his course.

394. Notwithstanding the copious directions which we have already given for the management of processes with the blowpipe, it may probably be as well, before we enter upon the description of the minerals, to make a few remarks upon the general manner of examining their habitudes.—Let it be supposed that the student has a mineral, which, in order to ascertain its name, he intends to examine by the blowpipe, and that this mineral has been properly prepared for that purpose, by the preliminary operations formerly described, the question then is,—how is he to begin?—In the first place, the mineral is to be heated in the matrass, in the manner, and with the objects, described in paragraphs 106, 138, 139. After which, it must be heated, alone, on charcoal or some other support, and the phenomena be observed which are described at paragraph 141; or, at least, such of them as may be developed. Lastly, the substance must be tried with the different fluxes, the various habitudes with each of which must be observed attentively. Paragraph 199 gives a tabular view of what may be expected to occur when a mineral is heated with a flux.

395. When an assay is exposed to the blowpipe, the heat that is first applied should be as low as possible; the temperature of the air on the surface of the jet of flame which is projected from the lamp, is sufficient to produce on many minerals a very sensible effect. Thus, in minerals which exhibit the phenomena of phosphorescence, that property is best elicited in this heat; the fusible inflammable minerals begin to melt; and various saline substances lose their water of crystallization. This temperature, therefore, is that in which the properties of minerals should be first investigated. To produce this heat most conveniently, the flame should be directed immediately above the assay, not

directed *towards* or *upon* it.—In the OUTER FLAME of the blowpipe jet, a mineral is raised to a tolerably full red heat, which produces the following effects. *Change of Colour :* thus, the yellow ores of iron become red, and the peach blossom tinge of flowers of cobalt becomes blue. *Earthy hydrates lose their water*, and then assume a particular appearance : thus, gypsum exfoliates, and prehnite and mesotype throw up coarse and irregular ramifications. Certain minerals begin to *tinge the flame ;* thus, carbonate of strontian produces its peculiar crimson colour, and muriate of copper its characteristic deep green flame. *The roasting of the metallic ores* is best carried on at this heat : sulphur and selenium are driven off, and are discovered by their peculiar odours ; arsenic is sometimes volatilized, but not so completely as by the inner flame; grey antimony melts; native bismuth exhudes from the pores of its matrix ; and pearlspar blackens and becomes magnetic.— The INTERIOR FLAME, which possesses a still higher degree of heat, produces another series of phenomena. These are much too numerous to be stated here, and it is, indeed, unnecessary to adduce examples; for the habitudes of mineral constituents described in the preceding pages, if considered in conjunction with the tables at paragraphs 139, 141, 180, will afford all the information on this point which the student will find necessary. It is sufficient to say here, with respect to the effects of the interior flame, that *fusion* and *reduction*, more or less perfect, are what are produced by it, and that the chief thing to be attended to, is the nature of the substance that is the result of the experiment.

396. " Most of the inflammable substances, when exposed to the apex of the flame, begin to liquefy, unless entangled in much earth, which yet does not always prevent their inflammation. When they are once inflamed, let the blast be stopped until they have burned away, either alone or with a flux ; and the residuum, if any, be examined afterwards by the flame." *Bergman.*—In trying the habitudes of the earthy minerals with the blowpipe, recourse to fluxes is seldom necessary ; but, when metallic substances are operated upon, fluxes are at almost all times a useful and necessary addition, since the colour derived by the latter from the former, is one of the principal characteristics by which the metals are recognized.—In all cases where a metallic globule is obtained, it should be separated from the adhering scoriæ by washing and pounding (see the reducing process), and examined with regard to its mallea-

bility and the other external properties by which the individual metals are distinguished. If the name of the metal is not discoverable by these means, it must be placed again before the blowpipe, and its habitudes, with and without fluxes, be compared with those which are described in the preceding pages, as characteristic of the different mineral constituents. By proceeding in this manner, the student cannot fail to identify the substance that is operated upon.

Class 1. COMBUSTIBLE MINERALS.

397. For the *Essential Characters* of this Class, see paragraph 389.

It is divided into two Orders, as follows :—

Order 1. *Combustible with Flame.*

Order 2. *Combustible without Flame.*

These Orders are not divided into Genera and Families ; because the species comprehended by each are few in number and easily discriminated.

Order 1. Combustible with Flame.

398. *Mineral Oil.*—Two substances are comprehended under this term; both of which are liquid, highly inflammable, and lighter than water :—

a. Naphtha. Perfectly fluid; transparent; nearly colourless. Sp. gr. 0·7.—Takes fire on the approach of flame; gives a bright blueish light, a considerable smoke, and a penetrating odour; leaves no residuum.

b. Petroleum. Fluid, but of a thicker consistence than common tar; viscid; translucent; dark brown; bituminous odour. Sp. gr. 0·9.—Burns with

a wick when heated ; during combustion, gives out
a very thick black smoke ; leaves a little black coaly
residuum.

399. *Mineral Pitch.*—There are three varieties
of this, which vary from solid to semi-fluid ; all
become soft and viscid when heated ; burn readily
with a bright yellow flame, leaving a small residue :—

a. Cohesive Mineral Pitch, Mineral Tar, Earthy
Bitumen, Maltha. Amorphous ; earthy ; brown ;
dull ; sectile ; very soft ; strong bituminous odour.
—Burns with a brisk clear flame, emits an agreeable
odour, and deposits much soot.

b. Elastic Mineral Pitch, Mineral Caoutchouc,
Elastic Bitumen. Nodulous masses ; greenish brown
and black ; translucent ; soft ; flexible ; elastic ; sectile ; strong bituminous odour. Sp. gr. 1·0.—Burns
with a large flame and much smoke. Melts in a
gentle heat.

c. Compact Mineral Pitch, Compact Bitumen.
Asphaltum. Massive ; brownish black ; fracture
conchoidal with a shining resinous lustre ; opaque ;
brittle ; has a bituminous odour, when rubbed.
Sp. gr. 1·6—1·0.—Burns freely, leaving a small quantity of ash.

400. *Brown Coal*, Wood Coal, Bovey Coal, Fibrous Coal, Bituminous Wood, Carbonated Wood,
Surturbrand.—Massive ; fracture earthy or fibrous,
with a woody structure ; colour blackish brown ;
soft. Sp. gr. 1·1.—This substance is principally
characterised by its burning with a weak flame,
and an odour resembling that of peat.

b. Kimmeridge Coal, Brown Bituminous Shale.
This is a variety of Brown Coal, having a fine slaty
structure, and a liver brown colour.—Burns feebly,
leaving a large earthy residue.

401. *Jet*, Compact Coal, Pitch Coal.—Colour
black ; lustre shining ; fracture perfectly conchoidal ;
occurs in branchiform masses ; structure ligneous ;

does not soil; soft. Sp. gr. 1·2.—Burns with a greenish flame and a strong bituminous odour, leaving a yellow ash.

402. *Black Coal,* Common Coal, Foliated Coal, Slate Coal.—Colour black; often with an iridescent tarnish; massive; structure thick slaty; fracture uneven; lustre shining resinous; soils. Sp. gr. 1·3.— Burns with a bright flame and much smoke.

403. *Cannel Coal,* Candle Coal, Parrot Coal of Scotland.—Colour greyish black; massive; fracture imperfectly slaty in the large; flat conchoidal in the small; glimmering; does not soil. Sp. gr. 1·2. —Burns with a bright flame, at the same time decrepitating and flying into angular fragments.

404. *Amber.*—Occurs in nodules; orange-colour, yellow, and yellowish white; fracture conchoidal; transparent; brittle; strongly resino-electric by friction; yields easily to the knife. Sp. gr. 1·1.— Burns with a yellow flame, a copious smoke, and flagrant odour, intumesces at the same time, but scarcely melts. The transparent kinds volatilize entirely. The opaque leave a residuum.

405. *Retinasphalt.*—Occurs in small irregular lumps; colour pale brownish yellow; opaque; glistening resinous; fracture conchoidal; brittle; soft. Sp. gr. 1·1.—When held to a candle, it melts, smokes, and burns with a bright flame and a fragrant odour.

406. *Fossil Copal,* Highgate Resin.—Amorphous; colour yellowish brown; translucent; lustre resinous; brittle; yields easily to the knife; rather heavier than water.—When heated, it gives out a resinous aromatic odour, and melts into a limpid fluid; when applied to the flame of a candle, it takes fire and burns with a clear yellow flame and much smoke, leaving no residuum.

407. *Dysodile.*—Massive; lamellar; yellow or grey; extremely fragile. Sp. gr. 1·5.—Burns with a considerable flame and smoke, and a crackling noise;

giving out an almost insupportably fœtid odour, and leaving a residue of nearly half its weight, unchanged in form.

408. *Sulphur*, Brimstone. — Occurs in nodular masses, or crystallized in acute octohedrons; colour yellow or orange, with a tinge of green; transparent or translucent; fracture conchoidal; lustre shining resinous; very brittle; remarkably resino-electric by friction. Sp. gr. 2·0.—Easily inflammable; burns with a lambent blue flame and a suffocating odour; fuses into a brown liquid.

b. Volcanic Sulphur. Occurs in the fissures of lava; generally stalactitic or pulverulent; agrees with the preceding in its other characters.

ORDER 2. Combustible without Flame.

409. *Mineral Charcoal*, Mineral Carbon.—Occurs in thin laminæ; greyish black; fibrous or wood-like in texture; friable; glimmering silky lustre; soils strongly.—Burns nearly as rapidly as common charcoal, with neither flame nor smoke.

410. *Blind Coal*, Glance Coal, Anthracite.—There are three varieties of this mineral, all of which burn without either flame or smoke, leaving a little ash :—

a. Massive Anthracite, Conchoidal Glance Coal. Iron black; often with a splendent metallic tarnish; fracture conchoidal; shining; light; brittle.

b. Slaty Anthracite, Slaty Glance Coal, Blind Coal, Stone Coal, Welch Culm, Kilkenny Coal. Brownish black; fracture slaty in one direction, conchoidal in the other; lustre inclining to metallic; easily frangible; sectile; brittle. Sp. gr. 1·6.

c. Columnar Anthracite, Columnar Glance Coal. Occurs in small short prismatic concretions, often curved; iron black; shining; often with a metallic tarnish; opaque; soft; light; brittle.

411. *Plumbago*, Graphite, Black-Lead.—Occurs in kidney-shaped masses; structure lamellar; colour iron grey; lustre glistening metallic; soft; unctuous; perfectly sectile; soils strongly, giving a distict mark on paper. Sp. gr. 2·1.—Before the blowpipe it is infusible, but, by a long-continued red heat, it burns without flame or smoke, leaving a portion of red oxide of iron. It is not acted on by the fluxes.

412. *Mellite*, Honeystone, Mellate of Alumina. —Occurs in grains and octohedral crystals; colour honey-yellow; translucent; shining; soft; brittle; refracts doubly. Sp. gr. 1·6.—In the matrass, alone, whitens, gives off water, and turns opaque. On charcoal, becomes of an opaque white with black spots, contracts, and at last is reduced to ashes, which slightly effervesce with acids.

413. *Diamond.*—Occurs crystallised, and in roundish grains which often present indications of crystalline faces; form, the regular octohedron and its varieties, which are usually curvilinear polyhedrons; structure perfectly lamellar, with joints parallel to the faces of the octohedron; in general colourless, but sometimes with a tinge of grey, blue, red, yellow, green, brown or black; lustre highly splendent, adamantine; more or less transparent; possesses a high degree of refractive power; hardness superior to that of every other substance. Sp. gr. 3·5.—Infusible before the blowpipe; but, at a white heat, it slowly burns away.

Class 2. METALLIC MINERALS.

414. For the *Essential Characters* of this Class, see paragraph 390.

It is divided into two Orders, as follows:—

ORDER 1. *Volatilizable, wholly or in part, on charcoal, before the Blowpipe, into a Vapour which condenses in a pulverulent form on a piece of Charcoal held over it.*

ORDER 2. *Fixed; not Volatilizable, except at a white heat.*

These Orders are respectively divided into Genera, in agreement with certain subordinate general characters.

ORDER 1. Volatilizable, wholly or in part, on charcoal, before the Blowpipe, into a Vapour which condenses in a pulverulent form on a piece of Charcoal held over it.

415. This Order is divided into three Genera, as follows:—

GENUS 1. *Entirely, or almost entirely, Volatilizable.*

GENUS 2. *Partly Volatilizable; the Residue affording Metallic Grains, with Borax, on Charcoal.*

GENUS 3. *Partly Volatilizable; the Residue not Reducible to the Metallic state with Borax.*

Every Genus is farther divided into Families.

GENUS 1. ENTIRELY, OR ALMOST ENTIRELY, VOLATILIZABLE.

416. This Genus is divided into two Families:—

FAMILY 1. *Lustre Metallic.*

FAMILY 2. *Lustre Non-Metallic.*

FAMILY 1. *Lustre Metallic*.

417. *Native Arsenic.*—Colour pale lead grey; externally dull; occurs botryoidal; structure concentric lamellar; fracture granular, often divergingly fibrous; soft; frangible. Sp. gr. 5·7.—Fuses; burns with a blueish flame; and gives off a dense white arsenical vapour; is all volatilized, except a small residue chiefly of iron, but mixed with silver or gold. In the matrass, metallic arsenic sublimes, and leaves a small bead of silver.

418. *Arsenical Native Antimony*. This is a variety of the Native Antimony described at paragraph 423, alloyed with a small portion of arsenic.—Alone, in the matrass, first gives off metallic arsenic, then fuses; heated red on charcoal, it is ignited, and gives off dense white arsenical fumes, which form round the metallic globule a net work of little crystals; by a prolonged blast the whole of the assay is dissipated in fumes.

419. *Native Bismuth.*—Colour reddish white, tarnished externally; occurs amorphous, and crystallised in octohedrons; structure perfectly lamellar; soft; not very frangible. Sp. gr. 9·0.—Fusible by the mere flame of a candle; before the blowpipe, fuses and is volatilized in the form of a white vapour, developing the odour of arsenic. In the open tube, gives off a little white arsenic. When cupelled, it tinges the bone ashes pure orange-yellow.

420. *Grey Antimony*, Sulphuret of Antimony.—Colour light lead grey, often iridescent; opaque; occurs crystallized, lamellar, and fibrous; lustre splendent; very soft; flexible; easily frangible. Sp. gr. 4·4.—Melts by the mere flame of a candle; evaporates almost totally before the blowpipe, in the form of a white vapour with a sulphurous odour.

421. *Molybdena.*—Colour lead grey; occurs in six-sided tabular crystals, and massive; structure lamellar; opaque; lustre shining; flexible, but

not elastic; unctuous; frangible; soils. Sp. gr. 4·6.
—Alone, on charcoal, scarcely fusible; gives a sulphurous odour; urged by a violent heat gives out
white vapours with a light blue flame; soluble in
carbonate of soda with violent effervescence.

422. *Native Tellurium*, Native Sylvan, Sylvanite.
—Colour tin white; shining; occurs in small crystalline grains, of a lamellar structure; soft; brittle.
Sp. gr. 6·1.—Fuses as easily as lead, then burns
with a light green flame, being almost entirely
volatilized in a dense white vapour, having a pungent acrid odour, sometimes like that of horse-radish, by which the presence of selenium is indicated.
In the open tube, Tellurium gives off a large quantity of fumes, which adhere to the sides of the tube
as a white powder, capable of fusion into clear colourless drops; a small portion escaping by sublimation.

423. *Native Antimony.*—Colour tin white, with a
yellow or black tarnish; occurs reniform and amorphous; structure lamellar; opaque; soft; frangible.
Sp. gr. 6·7.—Melts easily, on charcoal; gives a dense
grey inodorous vapour, and a melted button, which, if
cooled slowly, becomes covered with a net work of
brilliant white acicular crystals; generally a minute bead of silver remains.

424. *Native Quicksilver.*—Colour silver white; lustre
splendent; occurs in fluid metallic globules. Sp. gr.
13·6.—Volatilises entirely before the blowpipe, at
less than a red heat.

FAMILY 2. *Lustre Non-Metallic.*

425. *Cinnabar*, Sulphuret of Mercury.—Colour
carmine, cochineal red, or lead grey; streak florid
blood colour; occurs massive and crystallized; structure, lamellar, or fibrous; translucent. Sp. gr. 7·7.
—8·2.—On charcoal, fuses, and is volatilised with a
blue flame and sulphurous odour, leaving no residuum. In the open tube, it gives, by roasting, me-

tallic mercury, and a sublimate of cinnabar. In the
matrass, with soda, it produces globules of mercury.

426. *Red Antimony*, Sulphuretted Oxide of Antimony.—Colour cherry red, commonly with a blueish
or iridescent tarnish; occurs in diverging capillary
crystals, or amorphous; opaque; lustre shining;
brittle. Sp. gr. 4·5.—On charcoal, fuses readily, and
volatilizes with a sulphurous odour.

427. *Sulphuret of Arsenic.*—There are two varieties
of this; both of which, on charcoal, melt instantly,
and burn with a pale yellow flame, giving out sulphurous and arsenical vapours. In the open tube,
volatilize entirely, depositing white arsenic on the
upper part of the tube. In the matrass, fuse, boil
up, and give a transparent sublimate, of a dark yellow or fine red colour.—The following are the individual characters:—

a. Red Sulphuret of Arsenic, Realgar.—Colour
bright aurora red, scarlet, or orange; streak lemon
yellow; occurs massive, disseminated, or in prismatic crystals; lustre splendent, vitreous; soft; extremely frangible; acquires by friction resinous electricity. Sp. gr. 3·3.

b. Yellow Sulphuret of Arsenic, Orpiment.—Colour bright lemon or gold yellow; brilliant pseudo-
metallic lustre; occurs reniform and stalactitic; structure lamellar; in thin laminæ transparent and flexible. Sp. gr. 3·5.

428. *Bismuth Ochre*, Oxide of Bismuth.—Colour
greenish and yellowish grey; occurs massive; lamellar; shining or dull; opaque; soft; friable.
Sp. gr. 4·4.—Dissolves with effervescence in acids.
—On charcoal, is easily reduced to metallic globules,
which volatilize if the heat is prolonged. With borax, in the exterior flame, forms a colourless transparent glass.

429. *Antimonial Ochre*, Earthy Oxide of Antimony.—Colour yellowish brown; dull; earthy;
soft; brittle; heavy; occurs incrusting crystals of

grey antimony.—On charcoal, it whitens and evaporates without fusing; with borax, it intumesces and affords a few minute metallic globules.

430. *White Antimony*, Oxide of Antimony.—Colour whitish and greyish; occurs crystallized in tabular and acicular crystals, in diverging groups, and more rarely massive and investing; lustre shining, pearly; translucent; soft. Sp. gr. 5·5.—Alone, on charcoal fuses very easily, and sublimes in the form of a white vapour.

431. *Horn Quicksilver*, Muriate of Mercury, Corneous Mercury.—Colour greyish and greenish; occurs crystallized in quadrangular prisms, terminated by pyramids, also in tubercular crusts, rarely massive; lustre pseudo-metallic; translucent; soft; sectile. Dissolves in water, and gives an orange precipitate with lime water.—Alone, on charcoal, totally volatilizes with the odour of garlic, which intimates the presence of arsenic. With soda, in the matrass, affords many globules of mercury. With mic. salt and oxide of copper, on charcoal, it tinges the flame of a beautiful azure colour.

GENUS 2. PARTLY VOLATILIZABLE; THE RESIDUE AFFORDING METALLIC GRAINS, WITH BORAX, ON CHARCOAL.

432. This Genus is divided into two Families:—

FAMILY 1. *Lustre Metallic.*
FAMILY 2. *Lustre Non-Metallic.*

FAMILY 1. *Lustre Metallic.*

433. *Mispickel*, Arsenical Pyrites, Arsenical Iron.—Colour silvery white; occurs crystallized in right rhombic prisms, and amorphous; hard; brittle. Sp. gr. 6·5.—On charcoal, at first gives off dense and copious arsenical vapours, then fuses, exhaling the

odour of arsenic, into a brittle globule, resembling magnetic pyrites.

434. *Arsenical Cobalt*, White Cobalt Ore.—Colour silver white, often tarnished; occurs crystallized in cubes and octohedrons, also arborescent, stalactitic, and botryoidal; lustre glistening metallic; hard; brittle. Sp. gr. 7·7.—Alone, in the open tube, very readily gives off arsenious acid. In the matrass, some specimens yield a little metallic arsenic. On charcoal, exhales a copious arsenical vapour, and fuses partially and difficultly, into a grey metallic globule, which is not magnetic, and which continues brittle after long treatment with borax, to which it communicates a deep blue colour.

435. *White Copper*, Arsenical Copper Pyrites.— Colour yellowish white, generally tarnished; occurs amorphous; soft; brittle. Sp. gr. 4·5.—Yields a white arsenical vapour, and fuses into a greyish-black slag, which, upon being treated with borax, after a thorough roasting, yields metallic copper.

436. *Bright White Cobalt*, Cobalt Glance.—Colour silver or yellowish white: occurs crystallized in the cube and its varieties; also, arborescent, stalactitic, botryoidal, and amorphous; the planes of the cubical crystals are striated; structure lamellar; hard; brittle. Sp. gr. 6·4.—In the open tube, with a strong heat, gives off arsenious acid, exhales sulphurous acid, and whitens brazil-wood test paper placed in the tube. On charcoal, it first becomes black; gives off an abundance of arsenical fumes as it gets red hot; and, after some roasting, fuses into a dark coloured metallic globule, which attracts the magnet, and tinges borax deep blue.

437. *Arsenical Antimonial Silver*, Arsenical Silver Ore.—Colour silvery white on the fresh surface, tarnished blackish externally; occurs reniform; structure finely lamellar; glimmering metallic lustre; harder than antimonial silver; sectile; easily frangible. Sp. gr. 9·4.—Before the blowpipe, the anti-

mony and arsenic are mostly volatilized with a garlic
odour, leaving a globule of impure silver surrounded
by a slag.

438. *Copper Nickel*, Arsenical Nickel, Prismatic
Nickel Pyrites.—Copper red; occurs reticulated, dendritic, and botryoidal, but commonly massive; lustre
shining metallic; hard; brittle; difficultly frangible. Sp. gr. 6·6.—Dissolves in nitro-muriatic acid,
forming a grass green solution, which affords with
caustic soda or potass a pale green precipitate; whereas
from a solution of copper the precipitate is dark brown.
On charcoal, emits an arsenical vapour, and fuses
into a dark scoria mixed with white metallic grains.
In the open tube, the roasting produces a large
quantity of white arsenic, and leaves a greenish residuum, which, after farther roasting on charcoal,
yields by fusion with soda and borax a metallic globule of nickel, malleable, and very magnetic.

439. *Brittle Sulphuretted Silver*, Brittle Silver
Glance, Rhomboidal Silver Glance.—Colour dark
lead grey; occurs crystallized in low hexahedral
prisms, and massive; lustre varies from splendent to
dull; soft; easily frangible. Sp. gr. 5·9.—(Klaproth's analysis of this ore gave 10 per cent. of antimony, a substance of which Berzelius declares he
could perceive no trace.) Alone, in the open tube,
fuses and gives diminutive white arsenical crystals.
On charcoal, forms no deposit; yields the odour of
arsenic, and, after long roasting, parts with its sulphur. Aikin says that antimony flies off. Berzelius
asserts that it does not. The result of the fusion is,
a dull grey metallic globule, brittle, yet capable of
extension by the hammer. By a good oxidating
flame, the silver may be rendered pure.

440 *Silver Amalgam*, Argentiferous Mercury,
Native Amalgam.—Silver white; soft; sometimes
semifluid; also, crystallized in varieties of the rhombic dodecahedron; creaks when cut; whitens the
surface of copper when rubbed warm on it. S p.gr.

10·5.—In the matrass, intumesces, boils, gives off mercury, and affords a tumified mass, which, on charcoal, fuses into a globule of pure silver.

441. *White Silver*, Light Grey Silver Ore.—Light lead grey, often tarnished black ; occurs massive ; found mixed with galena ; fracture even or fibrous ; lustre glistening metallic ; soft ; frangible. Sp. gr. 5·6.—In the open tube, the odour of sulphurous acid, and a dense white antimonial vapour, are developed. After roasting, fusion ensues, and leaves a mass of scoriæ with which borax gives the tint of iron, and a globule of lead from which a little silver can be separated by cupellation.

442. *Antimonial Silver*, Antimoniated Native Silver.—Colour from silver to tin white, often with a red or yellow tarnish ; crystallized in prisms deeply longitudinally striated, also massive and in grains ; lustre shining metallic ; easily frangible ; slightly malleable ; soft. Sp. gr. 9·8.—In the open tube, dense white fumes of antimony fly off, and leave a particle surrounded by a ring of dark yellow glass. On charcoal, alone, gives off abundant fumes of oxide of antimony, and readily fuses into a grey, brittle, metallic globule, which after a prolonged blast, becomes pure silver. In the course of the process, the globule assumes a crystalline appearance, and a copious deposit of antimonial vapour is formed on the charcoal.

443. *Graphic Tellurium*, Graphic Gold, Graphic Ore, Graphic Gold Glance.—Colour steel grey ; lustre splendent ; occurs massive and crystallized in rhombic prisms ; yields to the knife ; brittle ; soils. Sp. gr. 5·7.—In the open tube, the tellurium is sublimed in copious white and grey fumes, which can be easily melted into liquid globules. A pungent odour is produced, but not like the putrid horse-radish odour of selenium. Alone, on charcoal, copious white fumes are developed that cover the support but disappear before the reducing flame, which they

colour green. A metallic globule is obtained, which gives out an intense heat at the instant of solidification; then becomes brilliant and malleable.

444. *Yellow Tellurium*, Yellow Sylvan Ore, Yellow Gold Glance.—Silver white and brass yellow; occurs in grains and minute rhombic crystals; lustre brilliant; soft. Sp. gr. 10·7.—Before the blowpipe, it behaves like the preceding.

445. *Bismuthic Silver.*—Colour lead grey; generally disseminated; lustre glistening metallic; soft; sectile; frangible.—Before the blowpipe, metallic globules ooze out, and form a mass on the addition of borax, to which flux it communicates an amber colour: subsequently, a tin-white, brittle, metallic button is formed.

446. *Sulphuret of Bismuth*, Bismuth Glance.— Light lead grey, or yellowish white; occurs in acicular crystals, sometimes radiating, often deeply striated longitudinally; also amorphous; soils; soft; brittle. Sp. gr. 6·1.—When held to the flame of a candle, it melts with a blue flame and sulphurous smell. Before the blowpipe, emits a reddish-yellow smoke which adheres to the charcoal; yields a sulphurous and sometimes an arsenious odour; the powder becomes white when it cools, but resumes the former colour when the flame is directed upon it; the residue of the assay is reduced to the metallic state with difficulty.—When melted with *borax*, bismuth may be distinctly precipitated by iron or manganese.

447. *Triple Sulphuret of Lead*, Bournonite, Cupreous Antimonial Sulphuret of Lead.—Blackish lead grey; lustre shining; occurs massive, and crystallized in rectangular prisms; structure perfectly lamellar; yields to the nail; very brittle; soils. Sp. gr. 5·8.—When thrown in powder on a hot iron, emits a blueish-white phosphorescent light. Alone, on charcoal, if suddenly heated, decrepitates, then fumes, emitting a dense, white, sulphurous vapour;

after which, there remains a globule of copper surrounded by a scoriaceous crust of sulphuret of lead.
After roasting the lead, a globule of copper may
be obtained with the aid of soda.

448. *Black Tellurism*, Nagyker Ore, Black Sylvan Ore.—Very dark lead grey; occurs in plates,
foliated; in thin laminæ flexible; soft; sectile; stains
a little. Sp. gr. 9·0. Soluble in acids with effervescence.—Alone, on charcoal, fuses and emits copious
fumes which partly condense on the support in the
form of a yellowish or reddish-brown powder;
this deposit can be dissipated by the interior flame,
which acquires a blue colour from it. By a strong
blast, a particle of malleable gold is obtained.

449. *Grey Copper*, Grey Sulphuret of Copper.—
yellowish or blackish grey; occurs massive and
crystallized in tetrahedrons; fracture conchoidal;
lustre shining; semihard; brittle. Sp. gr. 4·9.—
Deflagrates with nitre.—Before the blowpipe, decrepitates, and then melts into a brittle grey globule,
emitting a white vapour, with sometimes an arsenical odour. The fusion is assisted by borax. The
assay gives a yellowish or brownish red colour to
this flux, but does not unite with it.

FAMILY 2. *Lustre Non-Metallic.*

450. *Foliated Arseniate of Copper*, Copper Mica,
Rhomboidal Arseniate of Copper, Lamellated Hexahedral Arseniate of Copper.—Deep emerald green;
occurs massive, and crystallized in hexahedral laminæ, or tabular crystals, easily divisable; transparent; lustre semi-metallic; scratches gypsum but not
calc spar; sectile. Sp. gr. 2·5.—Before the blowpipe
alone, it decrepitates, gives off water and an arsenical
odour, passes to the state of a black spongy scoria,
and afterwards fuses into a black globule. With
borax, it affords a bead of copper.

451. *Lenticular Arseniate of Copper*, Obtuse
Octohedral Arseniate of Copper.—Sky-blue or bright

green; occurs in lenticular crystals; transparent; lustre vitreous; easily frangible; softer than. fluor spar. —Sp. gr. 2·9.—In the matrass, gives off much water; on charcoal, fuses, with an arsenical odour, into a black friable scoria, containing small white metallic globules.

452. *Acute Octohedral Arseniate of Copper.*— Brownish and yellowish green; crystallized in acute octohedrons, or capillary; transparent; lustre resino-vitreous; harder than fluor spar. Sp. gr. 4·2.—Intumesces, gives off arsenical vapours, fuses into a hard reddish brown scoria, which by subsequent fusion with borax affords a bead of copper.

453. *Hematitic Arseniate of Copper.*—Yellowish brown; occurs botryoidal, with a fine diverging fibrous structure; lustre silky. Sp. gr. 4·3.—Fuses into a hard black cellular scoria.

454. *Martial Arseniate of Copper.*—Sky blue; massive and in globular concretions, consisting of minute rhomboidal prisms; lustre shining vitreous; transparent; harder than calc. spar. Sp. gr. 3·4.— Fuses, gives out an arsenical vapour, and by subsequent fusion with borax yields a metallic globule.

455. *Arseniate of Lead.*—Wax yellow; occurs massive, acicular, and crystallized in hexahedral prisms; translucent; soft; sectile; frangible. Sp. gr. 6·4.—Alone, on charcoal, fuses difficultly, and is instantaneously reduced into numerous globules of lead, with copious disengagement of the vapour and odour of arsenic.

456. *Red Silver*, Ruby Silver, Sulphuretted Antimonial Silver.—Colour between cochineal red and lead grey; lustre shining; occurs crystallized in hexahedral prisms, massive, dendritic, and in membranes; streak cochineal red; soft; sectile; frangible. Sp. gr. 5·7.—Alone, on charcoal, slightly decrepitates, fuses with gentle effervescence, ignites, and gives off yellow and white sulphurous and antimonial vapours, (rarely also arsenical,) and

leaves a globule of silver. In the open tube, fumes much, exhales sulphurous acid. Vapour of oxide of antimony condenses abundantly in the tube, sometimes forming crystals, which may be wholly driven off by heat. After the assay has been some time exposed to the exterior flame, a button of pure silver results.

GENUS 3. PARTLY VOLATILIZABLE ; THE RESIDUE NOT REDUCIBLE TO THE METALLIC STATE WITH BORAX.

457. This Genus is divided into two Families :—

FAMILY 1. *Lustre Metallic.*
FAMILY 2. *Lustre Non-Metallic.*

FAMILY 1. *Lustre Metallic.*

458. *Common Iron Pyrites* (Arsenical Variety), Fer Sulphuré Arsenifère of Haüy.—Pale yellow, passing into steel grey ; crystallized, massive, and disseminated ; hard ; brittle. Sp. gr. 4·8.—Alone, on charcoal, fuses, exhales a vapour, having a strong odour both of sulphur and arsenic. Does not become magnetic.

459. *Tennantite.*—Massive, and in dodecahedral crystals variously modified ; externally tin white and splendent ; internally iron black and dull ; powder reddish grey ; structure lamellar ; brittle. Sp. gr. 4·4.—Alone before the blowpipe, on charcoal, burns, with slight decrepitation and a blue flame with the odour of sulphurous acid ; to which succeed copious arsenical fumes ; leaving a greyish black scoria which affects the magnet.

460. *Plumose Grey Antimony,* Capillary Sulphuret of Antimony, Feather Antimony.—Dark lead grey colour, often iridescent ; occurs in very minute capillary crystals, investing the surface of other

minerals as with a delicate down or wool, often so
interlaced and mutually adherent as to appear like
an amorphous crust; opaque; soft; brittle. Sp. gr.
3·5.—Fuses easily into a black slag, previously
giving out a vapour which condenses into a white
and yellow powder; the latter being sulphur, the
former oxide of antimony.

Family 2. *Lustre Non-Metallic.*

461. *Arseniate of Iron*, Cube Ore.—Olive green;
crystallized in cubes; lustre vitreous; streak straw
yellow; translucent; soft; brittle. Sp. gr. 3·0.—
Sometimes occurs decomposed, in the state of a
reddish-yellow powder.—Alone, in the matrass,
yields water and becomes red; if heated strongly,
slightly intumesces, but affords little or no white arsenic; the substance remaining, gives, when pounded,
a red powder. Alone, on charcoal, yields copious
arsenical fumes; fuses, in the reducing flame, into
a grey, metallic-brilliant, magnetic scoria; which, if
fused with fluxes, exhales the odour of arsenic, and
shows the presence of iron.

462. *Pharmacolite*, Arseniate of Lime, Arsenic
Bloom.—White, with a reddish or yellowish tinge;
occurs in delicate capillary crystals and botryoidal;
lustre glimmering silky; translucent on the edges;
soft; frangible. Sp. gr. 2·6.—Alone, in the matrass,
gives off much water, and becomes opaque; arsenious acid does not sublime. In the forceps, fuses,
in the exterior flame, into a white enamel. On
charcoal, in the interior flame, readily fuses into a
semi-transparent globule, yielding a copious arsenical
odour. If cobalt is present, the globule becomes
blue. With soda, decomposes, disengages much
arsenic, leaves lime on the charcoal.

463. *Oxide of Arsenic*, White Arsenic, Arsenious Acid.—Snow white, with sometimes a tinge of
red or yellow; occurs earthy and in capillary crystals; resembles Pharmacolite, but distinguished by

being soluble in water, which that substance is not.
—Before the blowpipe, alone, gives off the smell of
arsenic in the reducing flame, and completely volatilizes in the oxidating flame.   In the matrass,
sublimes without fusion, producing a crystalline
sublimate.

464. *Red Cobalt*, Arseniate of Cobalt, Cobalt
Bloom, Radiated Red Cobalt Ochre.—Peach-blossom
red; massive, and crystallized in short diverging
acicular prisms; shining; translucent; soft; rather
sectile; frangible. Sp. gr. 4·0.—Before the blowpipe
becomes grey, and emits an arsenical smell; Thomson says without smoke, Berzelius says with abundant fumes : tinges borax blue.

465. *Earthy Cobalt*, Cobalt Crust, Earthy Red Cobalt Ochre.—Peach-blossom red and blueish black;
massive and in velvety coatings; friable; dull; sectile; soft; streak ·shining; does not soil. Sp. gr.
2·2.—Alone, yields an arsenical odour: but does
not fuse; tinges borax deep blue and fuses with
it.

466. *Hepatic Quicksilver*, Hepatic Mercury, Hepatic Cinnabar, Quicksilver Liver Ore, Carbo-sul-
phuret of Mercury.—Of this, there are two varieties :—

a. Compact Hepatic Ore. Colour between dark
cochineal red and lead grey; amorphous; streak red
and shining; fracture even; opaque; glistening; soft;
sectile; frangible. Sp. gr. 7·1.—Alone, in the matrass, it gives cinnabar, and a black residuum.   The
cinnabar sublimes entirely, with a sulphurous odour;
the residuum, burnt in the open tube, gradually disappears, forming no deposit, but leaving a little
earthy ash.   This mineral is insoluble in nitric, and
soluble in muriatic acid.

b. Slaty Hepatic Ore. Colour as above, sometimes darker; principal fracture curved thick slaty,
shining; cross fracture uneven, glimmering; most
easily frangible; streak brownish. Sp. gr. rather less

than the above. In other respects the slaty and
compact agree.

ORDER 2. Fixed ; Not Volatilizable, except at a
white heat. .

467. This Order is divided into two Genera, as
follows :—

GENUS 1.   *Assume or Preserve the Metallic*
Form after Roasting on Charcoal, before the
Blowpipe, while any thing is dissipated, and
subsequent Fusion with Borax.

GENUS 2.   *Not reducible to the Metallic state,*
on Charcoal, before the Blowpipe, either with
or without Borax.

Each of these Genera is divided into Families.

GENUS 1. ASSUME OR PRESERVE THE METALLIC
FORM AFTER ROASTING ON CHARCOAL, BEFORE
THE BLOWPIPE, WHILE ANY THING IS DISSIPAT-
ED, AND SUBSEQUENT FUSION WITH BORAX.

468. This Genus is divided into two Families :—

FAMILY 1. *Lustre Metallic.*
FAMILY 2. *Lustre Non-Metallic.*

FAMILY 1. *Lustre Metallic.*

469. *Native Platinum,* Platina.—Colour between
steel grey and silver white ; occurs in small flat
grains ; hard ; malleable ; flexible. Sp. gr. 17·7.—Infusible ; incapable of oxidation ; has no action on
fluxes : is soluble in nitro-muriatic acid.

. 470. *Native Iron.*—Pale iron grey ; massive and in

thin plates; fracture hackly; hard; flexible; malleable; magnetic. Sp. gr. 6·5.—A rare mineral.

471. *Palladium.*—Occurs with Platina, in grains composed of diverging fibres; steel grey; lustre metallic shining; hard; malleable. Sp. gr. 12·1. —Infusible and unalterable before the blowpipe alone; melts with sulphur, which is volatilized by a continued heat, leaving a globule of pure malleable Palladium. This metal forms a deep red solution in nitric acid. A very rare mineral.

472. *Iridium* and *Osmium.*—These two metals occur forming a natural alloy, also alloyed with native Platina. Found in small flat grains; pale steel grey; shining metallic lustre; harder than Platina; brittle. Sp. gr. 19·6.—It acquires a dull black colour by fusion with nitre, but regains its proper colour and lustre when heated on charcoal. A very rare mineral.

473. *Native Gold.*—Bright yellow to orange yellow; in masses, grains, and crystallized in cubes and octohedrons; soft; ductile; flexible but not elastic; malleable; almost always alloyed with silver or copper; rarely with iron. Sp. gr. 15·0.—19·0., according to its purity.—Fusible into a button, but incapable of oxidation; has no action on fluxes.

474. *Native Silver.*—Pure silver white, generally tarnished greyish black; lustre shining; massive, capillary, reticulated, and crystallized in cubes and octohedrons; soft, but harder than gold; flexible; malleable. Sp. gr. 10·0.—Fusible into a globule, which is not altered by a continuance of the heat.

475. *Native Copper.*—Fine red colour, tarnished yellowish or blackish; crystallized in cubes and octohedrons, also massive, in plates, and threads; harder than silver; very tough; perfectly sectile; malleable; flexible. Sp. gr. 7·6.—8·6.—Fuses into a bead of apparently pure copper.

476. *Sulphuret of Silver,* Silver Glance, Vitreous Silver Ore.—Dark lead grey, often with an irides-

cent tarnish; occurs in masses, in threads, and crystallized in regular octohedrons and rhombic dodecahedrons; may be cut like lead; flexible; very malleable. Sp. gr. 7·0.—Alone, on charcoal, with a gentle heat, the sulphur evaporates; with a red heat, fuses, intumesces, and after some time collects into a globule, yielding a bead of silver surrounded by scoriæ, which, when fused with borax and salt of phosphorus, give traces of iron and copper.

477. *Black Silver*, Sooty Silver Ore, Argentiferous Variety of Grey Copper.—Dark grey; crystallised in tetrahedrons and massive; fracture small conchoidal; lustre shining metallic; hard; very brittle. Sp. gr. 6·2.—Readily fusible.

478. *Carbonate of Silver*, Grey Silver Ore.—Greyish-black; massive, and disseminated; streak bright; soft; brittle; heavy; effervesces with acids.—Easily fusible and reducible; exhales an antimonial vapour; froths with borax.

479. *Galena*, Sulphuret of Lead, Lead Glance.—This ore is divided into several subspecies; the two following are the principal:—

a. Common Galena.—Colour lead grey; occurs massive and crystallized in cubes and octohedrons, often with the angles truncated and edges replaced; fracture foliated; lustre splendent; fragments cubic; soft; sectile; easily frangible. Sp. gr. 7·6.—Alone, on charcoal, decrepitates, the sulphur is driven off, fusion ensues, and part of the assay sinks into the charcoal; ultimately, we obtain a globule of metallic lead. By cupellation, we ascertain whether this contains silver; we also learn from the colour of the bone ashes whether the lead was pure or not: if pure, the cupel is pale yellow, if it contain copper, it is greenish; if iron, blackish.

b. Compact Galena.—Massive, and in specular plates; fracture even; never in distinct concretions; fragments indeterminate; softer than common galena; streak brighter; agrees with it in other respects.

P

—Silver is generally present in galena; and, when the structure is fibrous, antimony.

480. *Blue Lead Ore.*—Blueish-grey; massive, and crystallized in six-sided prisms; surface rough; lustre glimmering; streak bright; fracture even; fragments indeterminate; soft; sectile; easily frangible. Sp. gr. 5·5.—Before the blowpipe, fuses with a low blue flame, emits a pungent sulphurous vapour, and is easily reduced to the metallic state.

481. *Yellow Copper Ore*, Copper Pyrites, Yellow Sulphuret of Copper.—Brass or gold yellow; often iridescent; crystallized in tetrahedrons with the angles and edges replaced, also, botryoidal, stalactitic, and amorphous; the amorphous variety is often beautifully variegated in the colour; yields to the knife; softer than iron pyrites, which it resembles in some respects; brittle. Sp. gr. 4·3.—Alone, on charcoal, emits a sulphurous vapour, and fuses into a brittle black globule, which, after a long exposure to the blast, affects the magnet. The action of this mineral with fluxes after roasting, is indicative of the presence of its ingredients,—iron and copper. With a small portion of borax, we obtain metallic copper. With soda, we obtain globules both of iron and copper, provided the roasting has been well performed. In the open tube, or in the matrass, no sublimate is given.

482. *Copper Glance*, Black Sulphuret of Copper, Black Copper Ore, Vitreous Copper.—Lead or iron grey; crystallized in hexagonal prisms, massive, and in granular distinct concretions; structure lamellar; fracture conchoidal or uneven; sectile; soft; frangible. Sp. gr. 5·6.—Alone, on charcoal, gives off a sulphurous odour, fuses easily in the exterior flame, without decrepitation, but with ebullition; affords a grey globule, sometimes magnetic: some species, while in fusion, exhibit a green pearl, which, on cooling, is covered with a brown crust, and which tinges borax green. In the open tube, the assay

burns, gives off a sulphurous odour, but forms no
sublimate. After roasting, the assay gives a globule of copper, with either soda or borax.

483. *Grey Copper Ore*, Fahl Ore, Grey Sulphuret of Copper.—Steel grey or iron black; powder
reddish black; crystallized in tetrahedrons, and massive; fracture imperfectly conchoidal; glistening;
semihard; brittle. Sp. gr. 4·5.—Phillips reckons
three sub-species of this mineral, the difference in the
composition of which is intimated by their names:—

1, Arsenical Grey Copper. Steel grey; crystallized; brittle.

2, Antimonial Grey Copper. Dark lead grey; not
regularly crystallized; not very brittle.

3, Platiniferous Grey Copper. A rare mineral.
With respect to the habitudes of these minerals
before the blowpipe, they may be divided into two
classes: those species which, in roasting, give off
arsenic, and those which give off antimony. Some
kinds decrepitate; others fuse, boil up, and fume,
all at once; others again, completely fuse, in the
first place, and then intumesce, forming excrescences
which have the appearance of a cauliflower in minia-
ture. When treated with soda, after roasting, they
all give a globule of metallic copper.

484. *Purple Copper Ore*, Variegated Copper Ore,
Buntkupfererz. — Colour between copper-red and
pinchbeck-brown, tarnished by exposure to air, red,
violet, azure, sky-blue, green,—sometimes these colours all tend to beautify a single specimen; massive,
in plates, and crystallized in cubes, the angles being
generally replaced, and the planes curvilinear; fracture rather conchoidal; shining; soft; somewhat
sectile; easily frangible. Sp. gr. 5·0.—Effervesces
with nitric acid, and tinges it green; deflagrates
with nitre.—Fuses readily, without smoke or smell,
but more quietly and less easily than Glance Copper, into a globule, which acts powerfully on the
magnetic needle. Tinges borax bright green.

FAMILY 2. *Lustre Non-Metallic.*

485. *Horn Silver*, Corneous Silver Ore, Muriate of Silver, Chloride of Silver.—Pearl grey, passing into greenish or reddish blue or brown, often tarnished; occurs crystallized in small cubes and acicular prisms, and amorphous; translucent; waxy lustre; very soft; ductile; flexible; malleable; easily frangible. Sp. gr. 4·8.—Melts in the flame of a candle.—Before the blowpipe, on charcoal, instantly melts, gives out muriatic vapours, and fuses into a bead whose colour, according to the purity of the assay, is pearl grey, brownish, or black and scoriaceous; when rubbed with a piece of moistened zinc, the surface becomes covered with a thin film of metallic silver. In the reducing flame, it is gradually converted into a globule of metallic silver.

486. *Red Copper*, Octahedral Red Copper Ore, Ruby Copper.—Divided, by Jameson, into four subspecies; which agree in the following chemical characters:—They easily melt and are reduced to the metallic state before the blowpipe on charcoal. If pulverised, and thrown into nitric acid, a violent effervescence ensues, and the copper is dissolved, the solution at the same time acquiring a green colour; but if thrown into muriatic acid, no effervescence takes place, although the ore dissolves. We can by this character distinguish red copper ore from red silver ore and cinnabar: red silver ore does not effervesce in nitric acid; and cinnabar does not dissolve in it. Red Copper is soluble in ammonia, to which it communicates a blue colour.

a. Foliated Red Copper Ore. Lead grey by reflected light, crimson red by transmitted light; massive, in concretions, and crystallized in varieties of the octohedron; lustre semi-metallic; fracture uneven, or imperfectly foilated; opaque when massive; semi-transparent when crystallized; yields to the knife; brittle; frangible. Sp. gr. 5·8.

b. Compact Red Copper Ore, Amorphous Red Copper. Colour between lead grey and dark cochineal-red; massive and spongiform; fracture even; opaque; yields to the knife; brittle; frangible. Sp. gr. 5·4.

c. Capillary Red Copper Ore. Carmine red; in capillary crystals and thin tables; lustre shining, adamantine; translucent.

d. Tile Ore, of which there are two kinds: *Earthy Tile Ore,* Ferruginous Red Copper, Earthy Red Oxide of Copper mixed with Brown Oxide of Iron.—Brick-red; massive and incrusting; composed of dull dusty particles more or less cohering; opaque; soils slightly; feels meagre; heavy.—*Indurated Tile Ore,* Indurated Brick-red Copper Ore.—Reddish and blackish brown; massive and in concretions; internal lustre resinous glimmering; soft; brittle; frangible. Sp. gr. 3·0.—Tile ore, before the blowpipe, becomes black, but is very difficultly fusible: it gives a dirty green colour to borax.

487. *Emerald Copper,* Dioptase, Silicate of Copper.—Emerald green; crystallized in elongated rhombic dodecahedrons; semi-transparent; lustre shining, pearly; scratches glass feebly; brittle; frangible. Sp. gr. 3·3.—Before the blowpipe, it becomes of a chesnut brown, and tinges the flame green, but is infusible; with borax it gives a bead of copper.

488. *Blue Copper Ore,* Blue Carbonate of Copper, Mountain Blue, Copper Azure.—Divided into two sub-species, which agree in the following chemical characters:—Soluble with effervescence in nitric acid, to which it gives a blue colour. Alone, in the matrass, gives off water, and blackens. Before the blowpipe, on charcoal, alone, blackens, but does not melt; with borax, it effervesces, gives a metallic globule, and colours the flux green:—

a. Earthy Blue Copper. Smalt blue; usually friable; seldom massive; soils a little. Sp. gr. 3·4.

b. Radiated Blue Copper. Azure blue; occurs in

small crystals, aggregated; also massive; lustre vitreous shining; translucent; soft; brittle. Sp. gr. 3·6.

489. *Malachite*, Green Carbonate of Copper.—Divided into two sub-species, both of which effervesce
with acids, and form a blue solution with ammonia.
Before the blowpipe, decrepitate and become black,
are partly infusible, and partly reduced to a black
slag.   They melt readily with borax, and afford
beads of copper, colouring the flux dark yellowish
green.

a. Fibrous Malachite.   Emerald green and grass
green; in concretions, slender fibres, and short capillary crystals; often stellated; fracture finely fibrous; fragments splintery; internal lustre shining
silky; translucent; soft; rather sectile; frangible.
Sp. gr. 3·7.

b. Compact Malachite, Massive Green Carbonate
of Copper.   Emerald green and whitish green, with
stripes; occurs massive, in various particular shapes,
commonly botryoidal or reniform; structure concentric lamellar in one direction, fibrous in the other;
fracture conchoidal or uneven; opaque; lustre glistening silky; harder than the preceding; brittle;
Sp. gr. 3·5.

490. *Muriate of Copper*, Copper Sand, Green Sand
of Peru.—Emerald green passing into olive; in the
form of sand, and crystallized in minute rhombic
prisms; structure lamellar; translucent; soft; shining; frangible. Sp. gr. 3·6.—When thrown on
burning coals, it communicates a green colour to the
flame.   It is soluble both in nitric and muriatic
acid without effervescence.   The solutions are green.
It tinges the flame of the blowpipe of a bright greenish blue; muriatic acid rises in vapours, and a bead
of copper remains on the charcoal.

491. *Phosphate of Copper.*—Externally greyish
black; internally verdigris green; massive and crystallised in short prisms; lustre glimmering; fracture fibrous diverging; opaque; streak apple green;

sectile; soft. Sp. gr. 3·5.—*Alone*, gives no colour
to the flame, but blackens and fuses; if a strong
heat is applied suddenly, it falls to powder; the
fused mass extends on the surface of the charcoal,
and acquires a reddish-grey metallic colour; in the
centre of the mass a small globule of metallic copper
is perceptible, which at the instant it congeals, emits
a very brilliant light.—With *borax*, this mineral behaves like pure oxide of copper. But a peculiar
phenomenon is presented when it is fused with soda.
If this flux is added in small successive portions, the
globule repeatedly liquifies till at last it become solid and infusible. When a great quantity of soda is
used at once, the saline mass is absorbed by the charcoal, and the copper is left on the surface.—According to Berzelius, the principal characteristic effect
of phosphate of copper is that produced by fusing it
with nearly an equal bulk of metallic lead. When
a mixture of this kind is exposed to a good reducing
flame, the whole of the copper separates in a metallic state, and there is simultaneously produced a
mass of fused phosphate of lead, which, upon cooling, crystallizes.

492. *Copper Green*, Chrysocolla.—Colour various
shades of green, often blueish; occurs botryoidal,
massive, and investing; fracture conchoidal; shining
resinous lustre; translucent; soft; brittle. Sp. gr.
2·2.—Before the blowpipe, it becomes first black,
then brown, but is infusible: on the addition of
borax, it melts rapidly, effervesces, then tinges the
flame green, and is reduced to the metallic state. In
diluted muriatic acid, it effervesces slightly; the
oxide of copper dissolves, and there remains behind
a nearly colourless and often semi-gelatinous mass of
silica, of the same size as the original specimen.

b. Pitch Copper.—A variety of the above mineral,
possessing a blackish brown colour, but agreeing
with it in other respects.

493. *Carbonate of Lead*, White Lead-Spar, White

Lead-Ore.—White, grey, and light brown; occurs crystallized in six-sided tables or prisms, fibrous, massive, and investing; fracture conchoidal or fibrous; lustre particularly adamantine; transparent; refracts doubly in a high degree; soft; very brittle. Sp. gr. 6·0.—7·2.—It is insoluble in water. It dissolves with effervescence in muriatic and nitric acid, especially if warm. Before the blowpipe, it decrepitates strongly, loses its white colour, becomes yellow, then red, and is immediately reduced to a metallic globule. It includes the following sub-species:—

a. Lead Grey, Carbonate of Lead. Colour grey; lustre metallic, but merely superficial.

b. Blue and Green Carbonate of Lead. Colour white, with spots and streaks of green and blue, caused by the infiltration of copper.

c. Earthy Carbonate of Lead. Grey, sometimes tinged greenish brown, massive and disseminated; fracture uneven; glistening, almost dull; soft; friable; heavy.

494. *Muriate of Lead*, Murio-Carbonate of Lead, Corneous Lead Ore, Chloro-Carbonate of Lead, Horn Lead.—Greyish or wine yellow; lustre splendent adamantine; crystallized in four-sided prisms, variously terminated; structure lamellar; transparent; very soft; sectile; easily frangible. Sp. gr. 6·0. —Alone, in the exterior flame, fuses into a transparent globule, which becomes yellow on cooling, and reticular externally; when again melted, it becomes white, and on increase of the heat, the acid flies off, and minute globules of lead remain behind. With oxide of copper dissolved in salt of phosphorus, a blue flame is produced around the assay globule, the usual effect of muriatic acid.

495. *Phosphate of Lead*, Green Lead Ore, Phosphorated Lead Ore.—Various shades of green and yellow; occurs crystallized in six-sided prisms, variously modified on the edges; also in particular shapes, often botryoidal; fracture uneven; lustre ada-

mantine; translucent; yields easily to the knife; brittle; frangible. Sp. gr. 6·4.—Before the blow-pipe, alone, on charcoal, in the exterior flame, usually decrepitates, then melts, and on cooling forms a dark-coloured polyhedral crystal, the faces of which present concentric polygons. In the interior flame, it exhales the vapour of lead; the flame assumes a blueish colour; and the globule on cooling forms crystals with broad facets, inclining to pearly whiteness. At the moment it crystallizes, a gleam of ignition may be seen in the globule. If the crystallized mass be pulverised and heated with borax, there results in the first place a milk white opaque enamel; upon the continuance of the heat, this effervesces, and at length becomes perfectly transparent, the lower part of it being studded with metallic lead. Cupellation elicits no silver from this lead.

496. *Brown Phosphate of Lead.*—Colour greyish brown; in other respects differs very little from the preceding mineral; occurs in minute six-sided prisms.—Before the blowpipe, melts into a globule, which concretes on cooling into a radiated mass.

497. *Sulphate of Lead,* Natural Vitriol of Lead.— Light grey and smoke grey; crystallized in rhombic prisms with diedral terminations; also massive; structure lamellar; fracture compact; lustre splendent adamantine; translucent; yields to the nail; brittle. Sp. gr. 6·3.—Decrepitates; in the exterior flame, on charcoal, fuses into a transparent glass, which becomes opaque as it cools. In the reducing flame, effervescence takes place, and a particle of lead is produced. When glass of soda and silica is used as a flux, the assay upon cooling assumes the colour of liver of sulphur.

498. *Chromate of Lead,* Red Lead-Ore, Red Lead-Spar.—Deep orange red; when pulverised, orange yellow; crystallized in oblique rhombic prisms; rarely massive; fracture uneven; lustre splendent adamantine; translucent; soft; frangible. Sp gr.

6·0.—Alone, on charcoal, decrepitates, and melts into a greyish slag, which spreads on the support; the lower part of the fused mass then developes the fumes peculiar to lead, and that metal is reduced; at the same time, the upper part of the mass gives a reddish brown powder. With borax, readily dissolves and produces a fine green glass, unless the chromate be in excess. With soda, the mass is absorbed by the charcoal, on the surface of which appear grains of metallic lead.

499. *Molybdate of Lead*, Yellow Lead-Ore.— Wax yellow and dirty honey brown; crystallized in octohedrons variously modified; rarely massive; fracture small conchoidal; lustre glistening resinous; translucent; soft; brittle. Sp. gr. 5·1.—Alone, decrepitates violently, and fuses into a dark grey mass, which sinks into the charcoal, leaving particles of reduced lead on its surface. If the absorbed part is washed, we obtain a mixture of malleable lead and unmalleable molybdena. With a limited quantity of borax a brownish glass is produced, with a large portion a blueish green glass. With mic. salt, a green glass is formed. With soda, fusion, absorption, and reduction, take place.

500. *Native Minium*, Variety of Oxide of Lead arising from the decomposition of Galena.—Vivid scarlet; amorphous and pulverulent; structure minutely crystalline.—Alone, first converted into oxide of lead, then reduced to metallic lead.

501. *Tinstone*, Oxide of Tin, Common Tin Stone. —Blackish brown, reddish brown, greenish white; crystallized in four-sided prisms, terminated by four-sided pyramids; also amorphous and nodulous; lustre splendent vitreous; fracture uneven; scarcely yields to the knife; brittle. Sp. gr. 6·7.—Alone, tin-stone strongly decrepitates. On charcoal, finely pulverised and mixed with borax, it may, by the long-continued action of the blowpipe, be reduced to the metallic state. When the dark-coloured va-

rieties are treated with soda, on platinum foil, they give traces of manganese. If columbium be present, the oxide of tin reduces very difficultly and very imperfectly, and when fused with a certain proportion of borax, the flux becomes opaque both by flaming and by cooling. The following varieties require to be noticed :—

a. Wood Tin, Fibrous Oxide of Tin. Brown and grey; in wedge-shaped detached pieces; divergingly fibrous; lustre silky.

b. Granular Tin, Stream Tin. Merely fragments and crystals of Tinstone, sometimes found separate, sometimes imbedded in a rock.

c. Columbiferous Oxide of Tin. Small crystals and grains imbedded in quartz; greyish red; lustre semi-metallic; opaque; hard. Sp. gr. 6·6.

502. *Nickel Ochre*, Oxide of Nickel.—Colour fine apple green; occurs as a thin coating, in loose powder, or friable; dull; meagre; light; stains; is insoluble in cold nitric acid.—Infusible, alone, but gives a hyacinth-red tinge to borax, by which it is reduced to the metallic state.

Genus 2. Not Reducible to the Metallic state before the Blowpipe, on Charcoal, either with or without Borax.

503. This Genus is divided into two Families:—

Family 1. *Magnetic after Roasting.*
Family 2. *Not Magnetic after Roasting.*

—

Family 1. *Magnetic after Roasting.*

504. *Magnetic Pyrites.*—Colour between bronze yellow and copper red; lustre shining metallic; amorphous, rarely crystallized; fracture somewhat conchoidal; hard; brittle; magnetic. Sp. gr. 4·5.— No change is produced upon it, in the matrass, alone; but, in the open tube, sulphurous acid is evolved,

but without the formation of any sublimate. On
charcoal, in the exterior flame, becomes red, and is
roasted into oxide of iron. In the interior flame,
fusion ensues, and a blackish crystalline mass, possessed of metallic brilliancy, is obtained.

505. *Common Pyrites*, Sulphuret of Iron, Magnetic Iron Stone, Martial Pyrites.—Brass and greenish
yellow, and steel grey; crystallised, amorphous, and
in every variety of particular shape; fracture granular uneven; brittle; too hard to yield to the knife,
which character serves at once to distinguish it from
copper pyrites which yields readily to the knife.
Sp. gr. 4·7.—On charcoal, its habitudes resemble
those of magnetic pyrites; but, tried in the matrass,
alone, we obtain, first a sublimate of sulphur, accompanied by the exhalation of sulphuretted hydrogen
gas, and afterwards a reddish sublimate similar to
sulphuret of arsenic. When well roasted, the assay
assumes a metallic aspect, and becomes attractable
by the magnet.

506. *White Pyrites*.—Tin white or greyish; crystallized in small octohedrons, botryoidal, and kidney form; hard; brittle; easily frangible. Sp. gr.
4·7.—Fuses, gives out a sulphurous vapour, and then
acts on the magnetic needle; it is much easier decomposed than common pyrites.

507. *Liver Pyrites*, Hepatic Pyrites.—Internal
colour steel grey; external liver brown; lustre glimmering metallic; occurs in most of the forms assumed by common pyrites, with which it agrees in
its other characters.

508. *Magnetic Iron Ore*, Native Loadstone or
Magnet, Oxydulated Iron, Octohedral Iron Ore.—
Iron black; lustre glimmering metallic; occurs
massive, and in varieties of the octohedron and
rhombic dodecahedron; fracture uneven; hard;
brittle. Sp. gr. 4·4.—It is magnetic, with polarity,
especially the massive variety. When heated, it becomes brown, but does not fuse: tinges borax dark

green.   The two following minerals are varieties of
this :—

b. Sandy Magnetic Iron-ore, Iron Sand, Granular Magnetic Iron Ore, Titaniferous Oxydulated
Iron.—Dark iron black; lustre brightly-shining
metallic; in small loose crystals or grains; fracture
conchoidal; brittle; yields to the knife. Sp. gr. 4·7.
—It is magnetic with polarity.—Before the blowpipe, alone, infusible and unalterable; but with
soda or borax, behaves like protoxide of iron. When
dissolved in salt of phosphorus, and completely reduced, it assumes, after the colour from the oxide
of iron has disappeared, a red colour, of various intensity, but deepest at the last instant of cooling.
The proportion of the titanium is indicated by the
intensity of the colour.

c. Earthy Magnetic Iron-Ore, Earthy Oxydulated
Iron.   Blueish black; massive; dull; soft; opaque;
soils; sectile; emits a faint clayey odour when
breathed on.   Sp. gr. 2·2.

509. *Iron Glance*, Red Iron-Ore.—Colour, when
finely divided, red; very feebly magnetic, in some
varieties scarcely at all so.—Infusible before the
blowpipe, on charcoal, alone, but becomes magnetic;
with borax, gives a dirty yellowish green glass.
There are the following varieties of this mineral :—

a. Regularly Crystallized Iron Glance, Specular
Iron Ore.   Dark steel grey; splendent metallic lustre; frequently tarnished, and beautifully iridescent;
massive; also crystallized in varieties of the rhomboid; fracture granular uneven; cherry red streak;
opaque; hard; brittle.   Sp. gr. 5·0.—When pulverised, it is slightly magnetic; but it does not, like
magnetic iron-stone, attract iron filings.

b. Irregularly Crystallized Iron Glance, Volcanic
Iron.   Occurs in very compressed and irregular crystals, often with curvilinear surfaces; external lustre
very bright; fracture conchoidal; shining; in other
characters agrees with variety *a.*

Q

c. Lamelliform Red Iron-Ore. Iron black; in straight or curved lamellæ; lustre shining metallic; the lamellæ in a strong light are translucent, and exhibit a blood colour; more highly magnetic than the other varieties.

d. Micaceous Iron Glance, Iron Mica, Micaceous Iron-Ore. Colour, by reflected light iron black, by transmitted light blood red; occurs in small six-sided scales or tables; loose or forming cells; sometimes massive; unctuous to the touch; when loose adheres to the fingers, but may be blown off without leaving any stain; thin kind translucent; streak cherry red; fracture foliated; brittle; exceedingly frangible. Sp. gr. 5·0.—Affects the magnet.

e. Red Scaly Iron-Ore, Red Iron Froth. Red with a tinge of brown; lustre glimmering semi-metallic; composed of scaly friable particles, which feel greasy and soil strongly; very soft; brittle; heavy.—Alone, before the blowpipe, infusible; but communicates to borax an olive or asparagus green colour. This mineral is rare.

f. Red Hæmatite, Kidney Iron Ore, Red Oxide of Iron, Fibrous Red Iron Ore. Colour between brownish red and steel grey; powder and streak red; internal lustre glimmering semi-metallic; friction produces a high metallic lustre; occurs in masses and every variety of forms,—stalactites, kidney-form balls, &c.; opaque; structure divergingly fibrous in one direction, and concentric lamellar in the other; fragments splintery; yields with difficulty to the knife; brittle. Sp. gr. 4·9.

g. Compact Red Iron Ore. Colour between brown red and steel grey; powder blood red; lustre shining; massive, slaty, and in pseudo-crystals; fracture even; yields to the knife. Sp. gr. 4·2.—When pure, not magnetic. Becomes darker before the blowpipe, but is infusible either alone or with glass of borax, to which, however, it communicates an olive green colour.

A. Red Ochre, Ochry Red Iron Ore. Light blood red with a tinge of brown; consists of faintly glimmering minute particles, often coating the other varieties, but sometimes found massive; nearly dull; friable; meagre; very soft; stains the fingers.—Sp. gr. 3·0.

δ. Reddle, Red Chalk. Colour light reddish brown; occurs massive, with an earthy fracture; streak lighter and more shining than the fracture; dull; opaque; soft; sectile; meagre to the touch; stains the fingers; writes easily; adheres to the tongue. Sp. gr. 3·5.

510. *Jaspery Iron Ore*, Jaspery Clay Iron-stone. —Reddish brown; massive; fracture conchoidal; lustre glimmering; fragments cubical; hard; brittle; frangible. Sp. gr. 3·2.—Blackens and becomes magnetic before the blowpipe.

511. *Brown Iron Ore.*—Colour, when powdered, blackish brown. Before the blowpipe, it blackens and becomes magnetic, but is infusible alone; it tinges borax olive green. In the matrass, all the hydrates of iron, under which term Berzelius includes both the brown and red iron ores, give off water, and leave a red oxide of iron. It gives a yellowish brown streak when rubbed on paper.—Aikin describes five varieties of this mineral, to which Phillips adds a sixth:—

a. Crystallized Brown Iron Ore. In cubes; brown; internally blueish grey; said to become magnetic when heated.

b. Scaly Brown Iron Ore. Colour between steel grey and clove brown; occurs in shining scales; semi-metallic; loose, or slightly aggregated; unctuous to the touch; stains strongly.

c. Brown Hæmatite, Fibrous Brown Iron Ore, Fibrous Hydrate of Iron. Clove brown or steel grey; occurs in various particular shapes; structure divergingly fibrous in one direction, concentric lamellar in the other; lustre silky; streak ochre yellow;

softer than red hæmatite; brittle. Sp. gr. 4·0.—Sometimes not magnetic.

 . *d.* Compact Brown Iron-ore, Compact Hydrate of Iron. Clove brown; occurs in masses of various and often very fantastic shapes; fracture flat conchoidal; lustre glimmering metallic; yields pretty easily to the knife; brittle; frangible. Sp. gr. 3·7.

e. Ochry Brown Iron Ore, Ochry Hydrate of Iron. Very light yellowish brown; coarse dull earthy particles; soft; friable; stains the fingers. When slightly heated, reddens.

f. Umber, Hydrate of Iron and Manganese. Clove brown; massive; fracture flat conchoidal; internal lustre resinous; very soft; sectile; easily frangible; soils strongly; feels meagre; adheres to the tongue; readily falls to pieces in water. Sp. gr. 2·2.

512. *Clay Ironstone,* Reniform or Kidney-shaped Brown Clay Ironstone, Oetites, Eaglestone, Nodular Ironstone.—Yellowish, blueish, and reddish brown; occurs massive, in globular, reniform, and various irregular balls, sometimes hollow, and formed of concentric layers; fracture earthy; glimmering; streak yellowish brown; sectile; frangible; meagre; yields easily to the knife. Sp. gr. 2·6.—Blackens and becomes very magnetic before the blowpipe.

513. *Columnar Clay Ironstone,* Columnar Red Clay Iron-ore, Scapiform Iron-ore.—Brownish red; occurs massive, and in globular and angular pieces composed of jointed columnar concretions, like starch; dull; rough; soft; brittle; exceedingly frangible; has a ringing sound; adheres to the tongue. Sp. gr. 3·4. Magnetic.—Becomes black before the blowpipe; bubbles up with borax, and communicates to it an olive green or blackish colour.

514. *Lenticular Clay Ironstone,* Lenticular Red Clay Iron Ore.—Reddish brown; in granular or lenticular distinct concretions aggregated into masses; streak red; soft; brittle; frangible. Sp. gr. 3·8.— Often magnetic.

515. *Pisiform Clay Ironstone*, Pea Iron Ore.—
Yellowish or blackish brown; in spheroidal grains the
size of peas, composed of curved concentric layers;
rough and dull externally, glistening internally; not
hollow; fracture fine earthy or even; streak yellowish brown; soft; brittle; frangible. Sp. gr. 3·1.—
Resembles Reniform Clay Ironstone in many respects, but differs from it in shape, and in not being
hollow as that mineral frequently is.

516. *Bog Iron Ore.*—There are three varieties of
this mineral. These are distinctly marked by their
different degrees of compactness, and the first is farther characterized by its very low specific gravity.
They are all hydrates of iron, and, therefore, according to Berzelius, give out water, when heated in the
matrass, and leave red oxide of iron. According to
Aikin, Bog Iron-ore becomes magnetic before the
blowpipe, and he seems to think, fusible.

a. Friable Bog Iron-ore, Morass-ore, Lowland
Iron-ore, Morassy Iron-ore. Pale yellowish brown;
dull dusty particles; friable; fracture earthy; soils
strongly; feels meagre; is light.

b. Indurated Bog Iron Ore, Swamp Ore, Swampy
Iron Ore. Dark yellowish brown; amorphous;
dull; fracture earthy; streak yellowish brown; very
soft; sectile; frangible. Sp. gr. 2·9.

c. Conchoidal Bog Iron-ore, Meadow Ore. On
the fresh fracture blackish brown; massive, in round
grains, and tuberose; shining; fracture conchoidal;
streak light grey; soft; brittle; frangible. Sp. gr. 2·6.

517. *Blue Iron Ore*, Phosphate of Iron.—This
species contains three varieties, all of which, according to Berzelius, behave before the blowpipe in the
following manner:—Alone, in the matrass, give off
much water, intumesce, and become spotted with
red and grey. On charcoal, intumesce, redden,
then fuse into a steel-coloured metallic globule.
With borax and salt of phosphorus, behave like
oxide of iron. With soda, on charcoal, in the re-

ducing flame, give grains of metal, which are attractable by the magnet. Afford no indication of manganese when tried on platinum foil.—Berzelius is
the only operator who says that metallic grains are
produced by the blowpipe. Aikin, Jameson, and
others, describe the product of the assay to be a
brownish black magnetic slag.

a. Foliated Blue Iron, Vivianite. Dark blue,
sometimes leek green ; in small acicular crystals,
deeply longitudinally striated ; splendent ; translucent ; as hard as gypsum ; flexible in thin pieces, but
not elastic ; fragments long tabular ; sectile ; frangible. Sp. gr. 2·8.

b. Fibrous Blue Iron. Indigo blue ; massive, in
roundish pieces, and in delicate fibrous concretions ;
glimmering internally ; opaque ; soft.

c. Earthy Blue Iron, Blue Martial Earth, Native
Prussian Blue.—Colour, when fresh dug, white, after
exposure to air, indigo blue ; occurs massive, disseminated, and investing ; very soft ; friable ; dull ;
meagre ; soils slightly ; rather light.—Dissolves readily in acids, but is rendered insoluble by exposure
to a red heat.

518. *Sparry Iron Ore,* Carbonate of Iron.—Pale
yellowish grey, passing when decomposing into brown
and black ; massive or in rhomboidal crystals ; internal lustre splendent pearly ; fragments rhomboidal ; structure lamellar ; translucent when light coloured ; yields easily to the knife ; not very brittle.
Sp. gr. 3·7.—In the matrass gives off no water.
Some species decrepitate violently. The assay blackens in a very gentle heat, and yields protoxide of
iron, strongly magnetic. Does not melt alone, but
dissolves with ebullition in glass of borax, which it
colours olive green.

519. *Black Iron Ore.*—There are two varieties of
this mineral :—

a. Fibrous Black Iron Ore, Black Hæmatite,
Black Ironstone, Fibrous Black Manganese Ore.

Blueish black, passing into steel grey; reniform and globular, exhibiting more or less of a fine and divergingly fibrous structure, with a glimmering imperfectly metallic lustre; gives a shining streak on paper; yields with some difficulty to the knife; is opaque and brittle. Sp. gr. 4·7.—Infusible alone, but affords with borax a violet-coloured glass.

b. Compact Black Iron Ore, Compact Black Manganese Ore, Black Ironstone, Black Hæmatitic Iron Ore. Differs from the preceding in having a conchoidal, or fine grained and uneven fracture. It occurs massive, and in distinct concretions, consisting of concentric lamellæ which are somewhat curved. In other characters, agrees with the preceding.

520. *Titanium.* The two following minerals are the only ores of this metal which are magnetic :—

521. *Menachanite*, Menaccanite.—Iron-black, passing to greyish; occurs in small grains like gunpowder, of no determinate shape; easily pulverised; powder attractable by the magnet; surface rough and glimmering; internal lustre semi-metallic; fragments sharp-edged; opaque; soft; brittle. Sp. gr. 4·4.—With two parts of fixed alkali, it melts into an olive-coloured mass, from which nitric acid precipitates a white powder. The mineral acids only extract from it a little iron. A mixture of the powder with diluted sulphuric acid evaporated to dryness, produces a blue-coloured mass.—Before the blowpipe, does not decrepitate nor melt. It tinges microcosmic salt green; but the colour becomes brown on cooling; yet microcosmic salt does not dissolve it. Soluble in borax, and alters its appearance in the same manner.

522. *Iserine.*—Iron-black, passing to brown, not altered in the streak; occurs in angular grains and rolled pieces; fracture conchoidal; brilliant semi-metallic lustre; opaque; scratches glass; attracts the magnet feebly. Sp. gr. 4·6.—Before the blowpipe, it melts into a blackish-brown glass, which

acts slightly on the magnet.   The mineral acids
have no effect upon it, but the acid of sugar extracts
a portion of Titanium.

523. *Octohedrite.*—This ore of Titanium belongs
properly to the next *Family,* among the minerals
composing which it will be found described.   We
merely insert the name here for the purpose of re-
marking that, according to the Comte De Bournon,
some of its crystals acquire by exposure to heat, the
property of acting slightly on the magnetic needle. ,

FAMILY 2. *Not Magnetic after Roasting.*

524. *Black Copper,* Black Oxide of Copper.—
Blueish or brownish black; disseminated; composed of dull dusty particles; soils slightly; shines in
the streak; heavy; forms a smalt blue coloured
solution with ammonia, the iron remaining undis-
solved.—Before the blowpipe, it emits a sulphurous
odour, and melts, alone, into a black slag; colours
borax green.

525. *Phosphate of Manganese,* Phosphate of Iron
and Manganese, Pitchy Iron Ore, Ferriferous Phosphate of Manganese, Phosphormangan.—Reddish-
brown, with a brilliant and somewhat chatoyant
lustre; massive; structure lamellar; opaque in the
mass; semi-transparent in splinters; scratches glass;
brittle; easily frangible. Sp. gr. 3·6.—Aikin places
this among the minerals which do not become magnetic after roasting; but Berzelius says it becomes
very magnetic. Phillips adopts Berzelius's assertion.
Jameson and Thomson, who both describe the habitudes of phosphate of manganese before the blowpipe, do not say that it becomes magnetic.   The
following is a summary of the experiments of Berzelius with this mineral.—Alone, in the matrass, a
little water is given off, which acts on test paper as
an acid.   If the assay is treated in the open tube,
held so that the flame may be directed into it, a deposition of silica takes place, which can only be

attributed to an evolution of fluoric acid. Alone, on charcoal, the assay fuses with great rapidity and a brisk intumescence into a black globule, which has a metallic lustre, and is *very magnetic.* With borax fuses easily; the oxidating flame producing the characteristic colour of oxide of manganese, the reducing flame, that of iron. With soda, on charcoal, fusion does *not* take place; but a large quantity of phosphuret of iron may be produced by the reducing experiment. If the assay be tried on platinum foil, the usual action of oxide of manganese takes place.

526. *Wolfram,* Tungstate of Iron.—Colour between dark-grey and brownish-black; frequently tarnished; occurs crystallized in rectangular parallelopipedons, in concentric lamellar concretions, and massive; lustre brilliant, often metallic; yields readily to the knife; brittle; frangible; the powder stains paper reddish-brown; easily separated into plates by percussion; moderately electric by communication; not magnetic. Sp. gr. 7·2.—As the pyrognostic characters of this mineral, as described by Aikin, Thomson, and Jameson, differ from each other very considerably, we give the experiments of Berzelius somewhat in detail.—Alone, in the matrass, decrepitates, and gives off water. On charcoal, may be fused by a good heat into a globule, the surface of which presents numerous iron grey coloured lamellar crystals, with a metallic lustre. Fusion readily takes place with borax, accompanied by the characteristic colour of iron. When fused with microcosmic salt, the glass assumes in the reducing flame a dark red colour; but, it is necessary to use an exceedingly minute portion of wolfram, otherwise the glass becomes opaque. If tin be fused with the assay, a green colour is produced; we must, however, take care that it be not too intense. In a good reducing flame, long kept up, the green colour is displaced by a faint red, and ultimately the wolfram is reduced.

When this mineral is tried with soda, on platinum foil, it decomposes and falls to powder, the edges of the soda being coloured green by the manganese it contains.—It is proper to remark, that wolfram frequently decrepitates very strongly, and splits into thin leaves; in these cases, the mineral will be found to have a hard yellow earthy coating, which is arseniate of iron, as may be proved by exposing the wolfram to the reducing flame, when a strong arsenical odour will be disengaged.

527. *Tin Pyrites*, Sulphuret of Tin, Bell-metal Ore.—Colour between steel grey and brass yellow; occurs amorphous; internal lustre shining metallic; fracture uneven; soft; brittle; frangible; not magnetic. Sp. gr. 4·3.—In the open tube, produces white fumes which do not volatilise, and developes the odour of sulphurous acid. Alone, on charcoal, melts easily, exhaling the odour of sulphurous acid, and in a high temperature, produces a white powder which covers the surface of the assay, and extends to a quarter of an inch round its base. This powder, the production of which is the principal pyrognostic character of tin pyrites, is an oxide of tin, and is not volatile either in the interior or exterior flame. After long roasting, a grey brittle metallic globule is produced, which gives with the fluxes the effects of copper and iron.

528. *Blende*, Sulphuret of Zinc, Black Jack, Zinc-Blende, Garnet-Blende, Dodecahedral Zinc-Blende, Pseudo-Galena. — Divided by Jameson into three species, viz.—Yellow, Brown, and Black Zinc-Blende.—One of the most distinguishing characters of this mineral is that it exhales a sulphurous odour when pulverised and digested in an acid. With respect to its behaviour before the blowpipe, the following description by Berzelius applies pretty well to all the species :—Alone, generally decrepitates with violence, but, even in a red heat, changes little in appearance. Does not fuse; in a very intense

heat, however, the thin places on the edges are a little rounded. Exhales a slight odour of sulphurous acid, but is difficult to roast. In the open tube, no change is experienced. On charcoal, a strong heating by the exterior flame, produces a circular deposit of vapour of zinc. The assay is very feebly attacked by soda; but by a powerful flame, the zinc is reduced, combustion ensues, and flowers of zinc are deposited on the charcoal.—The individual characters of this species are as follow:—

a. Yellow Zinc-Blende, Phosphorescent Blende. Lemon yellow, inclining to green; occurs in groups of mis-shapen octohedral crystals, massive, and disseminated; strong adamantine lustre; semi-transparent; refraction single; semi-hard; brittle; frangible. Sp. gr. 4·1.—Becomes phosphorescent by friction; and, according to Bergman, as powerfully under water as in the air.

b. Brown Zinc-Blende. There are two varieties of this; the *foliated* and the *fibrous*:—First, *Foliated Brown Zinc-Blende.* Reddish, yellowish, and blackish brown, sometimes tarnished with variegated colours; massive, in concretions, and crystallized in octohedrons, rhomboidal dodecahedrons, and tetrahedrons; shining externally; glimmering internally; fracture foliated; cleavage six-fold; translucent. Sp. gr. 4·0. not phosphorescent.—Second, *Fibrous Brown Zinc-Blende,* Cadmiferous Blende. Reddish-brown; massive; reniform, and in granular distinct concretions, intersected by curved lamellar distinct concretions; lustre splendent pearly; fracture fibrous; opaque. Sp. gr. 3·9.—This variety is remarkable, because it contains Cadmium as a constituent, whence it is termed by some Cadmiferous Blende. This discovery was made by Stromeyer, and Dr Clarke having remarked the great tendency which the salts of Cadmium have to crystallize, was led to suspect that all the fibrous ores of zinc owe their fibrous structure to the presence of Cadmium. This sagacious conjecture led him to examine various English ores of zinc, in all of which he found that metal. He thus describes the mode of detecting the presence of Cadmium in the carbonates and silicates of zinc: Place about the tenth of a grain of either of them in the state of powder on a slip of platinum foil ; apply the blowpipe flame ; upon this, if any oxide of Cadmium be present, it will be reduced and volatilized, and a protoxide will be deposited on the surface of the platinum of a peculiar reddish brown colour.

c. Black Zinc-Blende. Colour between greyish and velvet black, sometimes tarnished with variegated colours; when held between the eye and the light, appears blood-red, if transparent ; occurs in the same forms as Brown Zinc-Blende ; lustre semimetallic ; fracture foliated ; generally opaque; streak yellowish brown. Sp. gr. 4·0.

529. *Grey Manganese.*—There are five varieties of this mineral, which agree in the following characters:—Effervesce with muriatic acid, and give out chlorine.—Before the blowpipe, per se, on charcoal, in a good reducing flame, become brownish-red, but do not fuse. In borax, dissolve with a brisk effervescence arising from the disengaged oxygen, and tinge the glass violet.—The varieties follow :—

a. Fibrous Grey Manganese Ore.—Dark steel grey; occurs in very delicate acicular crystals, also in various particular shapes ; drusy surfaces ; shining ; streak dull and black ; soils strongly ; soft ; brittle.

b. Radiated Grey Manganese Ore, Striated Grey Ore of Manganese.—Steel grey, varying in intensity, and sometimes tarnished ; occurs crystallized in oblique four-sided prisms, also in distinct concretions, which are stellular, radiated, and collected into wedge-shaped masses ; lustre metallic ; cleavage prismatic ; rather difficultly frangible. Sp. gr. 4·6. —Agrees with the *first* species in other respects.

c. Foliated Grey Manganese Ore.—Steel grey, verging to iron black ; massive, disseminated, in

concretions, and crystallized in short oblique four-
sided prisms; lustre shining metallic; cleavage prismatic; fracture uneven; fragments blunt-edged;
streak dull and black; soils; soft; brittle; easily
frangible. Sp. gr. 3·7.

d. Compact Grey Manganese Ore.—Iron black;
massive; internal lustre glimmering metallic; fragments sharp-edged; dull and darker in the streak;
soils; soft; brittle; easily frangible. Sp. gr. 4·4.

e. Earthy Grey Manganese Ore, Ochre of Manganese, Wad.—Iron black, inclining to blue; massive, disseminated, and dendritic; friable; composed
of feebly glimmering semi-metallic fine scaly particles, which soil strongly, and more or less cohere.
In the matrass, gives off water abundantly; afterwards behaves before the blowpipe like the other
varieties.

530. *Sulphuret of Manganese,* Prismatic Manganese
Blende, Sulphuretted Oxide of Manganese.—Colour
when fresh broken dark steel-grey; after exposure
brownish black; massive and crystallized in oblique
four-sided prisms; lustre of the fresh fracture splendent, of the tarnished surface shining, and metallic;
fracture foliated, with different cleavages; opaque;
moderately hard; sectile; streak leek green. Sp.
gr. 3·9.—Alone, in the matrass, it remains unaltered.
Roasts in the open tube with tardiness and difficulty,
giving no sublimate. On charcoal, after being roasted by a powerful reducing flame to a certain extent,
the thin portions of the assay may be melted into a
brown coloured scoria. During the roasting, sulphur
is given out. When perfectly roasted, its habitudes
with the fluxes, are the same as those of pure oxide
of manganese. Colours borax violet, but not until
the assay is completely dissolved; in the meantime,
if the glass is suffered to cool, it shows a yellowish
tint. With microcosmic salt, fusion takes place, accompanied by a brisk effervescence, and the disengagement of a large quantity of gas, which continues

R

even when the blast is ended.   When the globule is
brought near the flame of the lamp, little detona-
tions are heard, which are occasioned by the combustion of the bubbles of liberated gas.   If the assay
globule be large, so as to retain a considerable quantity of heat, this curious phenomenon is prolonged.
At length, a large bubble of gas bursts out, the
inflammation of which is attended with a pale green
light.   During this experiment, the glass is remarkable for a very peculiar play of colours.   It is transparent and colourless while liquid, but on cooling
acquires a tinge of yellow.   When but a small part
of the mineral is decomposed, the mass as it gets
cold becomes first yellow, then brown, and at last
black; the production of the last colour resulting
from the separation of a fixed substance at the
moment of congelation.   If we examine the yellow
glass with a microscope, we perceive small black
particles suspended within it.   When the whole of
the sulphuret is dissolved, and there is a cessation of
effervescence, a glass is producedw hich is both colourless and pellucid, but which acquires in the
oxidating flame, a fine amethyst colour.—Tried with
soda, only a partial fusion of this mineral can be
effected ; there is formed a hepatic mass, which
sinks into the charcoal, leaving a half-fused grey
scoria on the surface.

531. *Earthy Cobalt*, Black Oxide of Cobalt, Black
Cobalt Ochre.—There are two species of this mineral, both of which behave before the blowpipe as
follows :—Alone, in the matrass, give off empyreumatic water ; on charcoal, exhale a white arsenical
vapour, but do not melt ; with borax and microcosmic salt, dissolve and give a deep cobalt blue colour ;
with soda they are infusible ; on the platinum
wire, a mass is produced coloured deep green by
manganese :—

a. Earthy Black Cobalt Ochre. Blueish or brownish
black, composed of dull dusty particles which soil

very little ; streak shining ; meagre ; friable ; light,
almost swimming in water.

b. Indurated Black Cobalt-Ochre. Blueish black ;
massive and coating ; glimmering externally ; opaque ;
fracture fine earthy ; streak shining ; very soft ; rather
sectile ; easily frangible. Sp. gr. 2·2.

532. *Tantalite*, Columbite, Prismatic Tantalum
Ore.—Blueish grey, verging to iron black ; occurs in
imbedded single octohedral crystals of the size of a
hazel nut ; lustre glimmering metallic ; opaque ;
fracture imperfectly foliated ; gives sparks with steel ;
not magnetic. Sp. gr. 7·9.—Alone, before the blowpipe, it suffers no change ; with borax, it dissolves
slowly, but perfectly. Berzelius has given us, in his
work on the blowpipe, a series of experiments on
columbite from different places, wherein he exhibits
the methods of reducing the different constituents of
this mineral, and of showing the variation in their
proportions.

533. *Yttrotantalite*, Yttrocolumbite.—Dark iron-
grey, yellowish brown, and dark brown ; occurs in
nodules the size of a hazel nut, and in rhombic
prisms ; powder grey ; opaque ; fracture foliated ;
lustre shining metallic ; scratches glass. Sp. gr. 5·5.
—Alone, in the matrass, gives off water ; the black
variety becomes yellow, and some specimens become
spotted. In a red heat, the assay whitens, and the
glass of the matrass is acted on ; the water which is
given out meanwhile first turns brazil-wood paper
yellow, and afterwards bleaches it. Fuses with
borax, into a colourless glass, which, when saturated,
becomes opaque. With microcosmic salt, the black
variety yields a glass, which, after being exposed to
a good reducing flame, acquires on cooling a slight
rose colour. The dark and yellow varieties, assume
on cooling a fine pale green tint. The rose colour is
indicative of columbium, the green of uranium. By
a trial on platinum foil, the presence of manganese
is shown.

534. *Titanium.*—The magnetic ores of this metal are included in the preceding *family*. The four following minerals are the ores of titanium which are not magnetic.

535. *Octohedrite*, Anatase, Octohedral Titanite, Pyramidal Titanium-ore.—Colour, by reflected light, blue of various shades, reddish brown, or steel grey ; by transmitted light, greenish yellow ; occurs in small octohedral or pyramidal crystals, the faces of which are transversely striated ; lustre splendent semi-metallic ; translucent ; scratches glass ; is easily broken. Sp. gr. 3·7. Many of the yellowish-brown crystals acquire by an exposure to heat, a deep blue colour.—Before the blowpipe, this mineral behaves like perfectly pure oxide of titanium. Alone, it is infusible ; mixed with an equal weight of borax, it melts into an emerald green glass, which, as it cools, crystallizes in needles ; by a farther addition of borax, the result is a clear glass of a hyacinth red colour ; if this is brought to the extremity of the flame, the colour changes to deep blue, and the glass becomes opaque ; but if the action of the blowpipe be continued, the blue becomes white, and at a still higher temperature, the original hyacinth red colour is restored ; thus, as the degree of heat is altered, the appearance and disappearance of the colours can be produced. In general, the native oxides of titanium dissolve but partially and with difficulty in salt of phosphorus, and the portion that does not melt becomes white and semi-transparent, and has the appearance of a salt mixed with the mineral.

536. *Sphene*, Rutilite, Calcareous Silicate of Titanium, Brown Ore, Prismatic Titanium-ore.—There are two varieties of this mineral, the habitudes of which before the blowpipe, according to Berzelius, are as follows :—Alone, in the matrass, give off a little water. The brown variety becomes yellow, but does not lose its transparency.—The yellow sphene suffers no change. A specimen from Fin-

land, presented a curious phenomenon : having been heated in a matrass till it began to melt, at a certain moment the assay suddenly shone as if it had taken fire. On charcoal, or in the forceps, sphene slightly intumesces, and fuses on the edges into a dark-coloured glass. The unfused portion preserves both its semi-transparency and its yellow colour. With borax it is easily fused into a clear transparent yellow glass, which becomes brown upon the addition of more sphene, but does not develope in the reducing flame the colour which is characteristic of titanium. A partial fusion is obtained with salt of phosphorus, the unfused portion becoming milk white. With this flux, in a good reducing flame, it is possible to develope the characteristic colour, especially if tin be added. The glass, after a long blast, becomes opaline on cooling. With soda, sphene produces an opaque glass, which no addition of the flux can render transparent, and which when cold is similar to that formed with pure oxide of titanium. If a large quantity of soda be used, the charcoal absorbs the glass, and by the reducing operation a little iron is usually produced.

a. Common Sphene.—Colour various shades of brown, green, and white ; occurs in granular distinct concretions, and crystallized in oblique rhombic prisms, modified ; internal lustre shining adamantine ; cleavage imperfect ; fracture imperfect conchoidal ; opaque to translucent ; streak yellowish white ; hard ; brittle ; frangible. Sp. gr. 3·5.

b. Foliated Sphene.—Colour various shades of yellow, sometimes passing into clove brown ; occurs massive ; in straight lamellar concretions ; and crystallized the same as the first variety ; lustre resinous, splendent on the cleavage, glistening on the fracture ; has a double cleavage, in which the foliæ are parallel with the lateral planes of the oblique four-sided prism ; in other characters agrees with the preceding variety.

537. *Titanite,* Rutil, Native Oxide of Titanium, Red Schorl.——Colour dark blood-red, passing into light brownish red; occurs massive, in distinct concretions, and in prismatic crystals, the surfaces of which are longitudinally striated; external lustre shining; structure lamellar; principal fracture foliated; cross fracture conchoidal; fragments cubical; translucent; hard; brittle; frangible. Sp. gr. 4·2. —" Before the blowpipe, it does not melt but becomes opaque and brown. With microcosmic salt it forms a globule of glass, which appears black; but its fragments are violet. With borax it forms a deep yellow glass, with a tint of brown. With soda it divides and mixes, but does not form a transparent glass."—*Thomson.* " Behaves like oxide of titanium, but the hyacinth colour it gives in the oxidating flame is never so pure as that given by anatase. Treated with soda on platinum foil, it colours the edges of the flux green,—a proof of the presence of manganese."—*Berzelius.*

538. *Nigrine,* Granular Oxide of Titanium.—Brownish black; occurs in loose, angular, and rounded grains; structure straight lamellar; fracture flat and imperfectly conchoidal; lustre glistening adamantine; opaque; brittle; streak yellowish brown. Sp. gr. 4·6.——Infusible before the blowpipe without addition; with borax, it melts into a transparent hyacinth-red globule.

539. *Pitchblende,* Protoxide of Uranium, Pitch Ore, Pechblende, Sulphurated Uranite.—Greyish, greenish, or brownish black; internal lustre shining resinous; massive, disseminated and in thick curved lamellar concretions; fracture conchoidal; streak black and shining; opaque; soft; very brittle. Sp. gr. 6·5.——Imperfectly soluble in sulphuric and muriatic acids, perfectly in nitric and nitro-muriatic acids; colour of the solution wine yellow. Infusible with alkalies in a crucible.—Before the blowpipe, alone, infusible and unalterable; however, if

held in the forceps, it colours the external flame
green. With borax, it forms a grey slag; with
microcosmic salt, a clear green globule; with soda,
does not fuse, but, if reduction be attempted, it gives
white metallic globules of iron and lead.

540. *Uranite*, Uran-Mica, Micaceous Uranitic
Ore, Chalcolite, Phosphate of Uranium, Hydrate of
Uranium.—Emerald green, apple green, and lemon
yellow; crystallized in rectangular prisms and in
tables; lustre splendent pearly; translucent; streak
green; fracture foliated; soft; sectile; easily frangible. Sp. gr. 3·3.—Dissolves without effervescence
in nitric acid, forming a lemon yellow solution.—It
decrepitates violently before the blowpipe on charcoal, is greatly diminished in weight by ignition,
and acquires a brass colour; with borax it yields a
yellowish green glass.

541. *Uran Ochre.*—There are two species of this
mineral, viz.

a. Friable Uran Ochre, Pulverulent Oxide of
Uranium.—Lemon yellow; usually coating pitch
ore; friable; composed of dull dusty particles; soils
feebly; light.—In the matrass yields water and
possesses a red colour while hot; in the reducing
flame assumes a green colour without melting; behaves like pure oxide of uranium in other respects.

b. Indurated Uran Ochre, Compact Oxide of
Uranium.—Colour sometimes lemon yellow, sometimes reddish and brownish; massive and disseminated; opaque; glimmering; fracture imperfect
conchoidal; soils; soft; sectile. Sp. gr. 3·2.—Alone
in the matrass, yields water and becomes red. On
charcoal in a good heat, fuses into a black globule.
Behaves like the preceding variety with fluxes.

542. *Cerite*, Cerium Ore.—Colour between crimson red and clove brown, sometimes flesh red;
powder grey; massive and disseminated; glimmering resinous internally; semi-transparent; streak
greyish white; fracture fine splintery; semi-hard;

brittle. Sp. gr. 4·6.—Alone in the matrass, yields
water and becomes opaque; on charcoal, it splits into pieces but does not melt; when powdered, the
colour changes from grey to yellow; with borax, a
solution is slowly effected. The oxidating flame
produces a deep orange coloured glass, which on
cooling becomes of a clear yellow; if we resort to
the process of *flaming*, the glass acquires the appearance of a white enamel.

543. *White Manganese*, Carbonate of Manganese.
—White, yellowish, or reddish; massive and crystallized in curvilinear rhomboids; translucent;
glistening lustre; scratches glass; is brittle. Sp.
gr. 2·8.—It gives off a little water, and violently decrepitates, when heated alone in the matrass. In
the reducing flame, on charcoal, it becomes brownish
black. It is infusible per se, but tinges borax purple. It effervesces with acids, and on digestion with
muriatic acid, gives out chlorine.

544. *Calamine*, Rhomboidal Calamine, Carbonate
of Zinc. There are four varieties of this:—

a. Sparry Rhomboidal Calamine.—Greyish and
yellowish; in small rhomboidal crystals, and in various particular shapes; shining and pearly internally;
partially transparent; yields easily to the knife.
Sp. gr. 4·3.—Dissolves with effervescence in muriatic acid. Before the blowpipe it assumes the appearance of white enamel, and then is acted upon
as if it were pure oxide of zinc. When calamine
which contains cadmium, is exposed on charcoal to
the reducing flame, the first impulse of the heat produces around the assay a red or orange coloured ring.

b. Compact Rhomboidal Calamine, Lapis Cala-
minaris.—Greyish, greenish, yellowish, or brownish;
massive, in concretions, and other shapes; rarely in
crystals; dull internally; fracture splintery; nearly
opaque.—Chemical characters same as variety *a.*

c. Earthy Rhomboidal Calamine, Sub-Carbonate
of Zinc.—Greyish white; dull; massive and invest-

ing; fracture earthy; yields to the nail; opaque; adheres to the tongue. Sp. gr. 3·6.—Tried in the matrass alone, it gives off water, and afterwards behaves like oxide of zinc. The assay may be volatilized by a continued blast in the reducing flame, when it leaves a scoria containing iron.

d. Cupriferous Calamine, Double Carbonate of Zinc and Copper.—Thin crystalline laminæ, divergingly aggregated, and with a silky lustre.—In the matrass alone, fumes and becomes black, produces with fluxes the effects of copper. By the reducing experiment with soda, the charcoal is covered with fumes of zinc, and a metallic globule of copper is produced.

545. *Electric Calamine,* Prismatic Calamine, Silicate of Zinc.—Greyish and yellowish white; massive, in particular shapes, and sometimes in rhombic prismatic crystals; internal lustre glimmering; fracture small uneven; transparent to opaque; yields to the knife. Sp. gr. 3·4.—When gently heated, is strongly electric; forms a solution in muriatic acid, without effervescence, which gelatinizes on cooling. Alone in the matrass, it slightly decrepitates, gives off water, and becomes white; in a strong heat, intumesces a little, but does not fuse. Gives with borax a colourless glass which is permanently transparent. It swells up with soda, but does not fuse. The solution of cobalt produces with it, first a green colour, and afterwards, in a good heat, a most beautiful blue.

546. *Tungsten,* Tungstate of Lime.—Greyish or yellowish white; amorphous, in granular concretions, and crystallized in octohedrons; fracture conchoidal; lustre adamantine; translucent; yields to the knife. Sp. gr. 6·0.—Before the blowpipe it crackles and becomes opaque, but does not fuse. —With borax it forms a colourless glass, unless the flux be in too great a quantity, when the glass is brown. With microcosmic salt it forms

a blue glass, which loses its colour in the exterior flame, but recovers it in the interior flame.

547. *Chromate of Iron,* Prismatic Chrome-ore.—Brownish black; formed massive and in octahedrons; fracture conchoidal; lustre between resinous and metallic, shining; scratches glass; opaque; brittle. Sp. gr. 4·4.—Fused with potass and dissolved in water, it forms a beautiful orange-yellow coloured solution. Insoluble in nitric acid. Not magnetic. Does not fuse before the blowpipe per se, but forms a fine green bead with borax, having a different appearance from the green mass produced by fusing borax with magnetic iron-ore.

CLASS 3. EARTHY MINERALS.

548. For the *Essential Characters* of this Class, see paragraph 391.

It is divided into three Orders, as follows :—

ORDER 1. *Soluble, either wholly or in considerable proportion, in cold and moderately dilute Muriatic Acid.*

ORDER 2. *Fusible before the Blowpipe, alone.*

ORDER 3. *Infusible before the Blowpipe, alone.*

These Orders are divided into Genera, according to certain subordinate characteristics.

ORDER 1. Soluble, either wholly or in considerable proportion, in cold and moderately dilute Muriatic Acid.

549. This Order is divided by Aikin into two

Genera: the first comprising those minerals which
" effervesce vigorously," the second, those which
" effervesce very feebly in cold, but more vigorously
in warm, muriatic acid." But, as the last Genus
comprehends only two species, namely, Pearl Spar
and Dolomite, the distinction has not been adopted.

550. *Carbonate of Lime.*—No other mineral can
be compared with Carbonate of Lime in the abundance with which it is scattered over the earth.
Many mountains consist of it entirely, and hardly
a country is to be found on the face of the globe,
where, under the names of limestone, chalk, marble,
spar, it does not constitute a greater or smaller part
of the mineral riches. The sub-species may be
reckoned in number 14. These differ greatly in
external characters; but all agree in these two respects: they readily yield to the knife; their specific
gravity is below 3·0.—Carbonate of Lime, in all its
varieties, has the following chemical characters:—
Alone, in the matrass, yields no moisture. When
pure, on charcoal, is infusible, per se, and does not
tinge the flame of the blowpipe, but it shines with
peculiar brightness when the carbonic acid is driven
off. It loses, by complete calcination, about 43 per
cent.; becomes caustic thereby, heats with water,
and acts as an alkali on reddened litmus paper.
Ferruginous or manganesian calcareous spar blackens
by heat. With the fluxes, dissolves with effervescence, and behaves like Lime. Those varieties, which
contain iron or manganese, give a coloured glass.
    a. Calcareous Spar, Calc Spar. Colour very vari-
ous; occurs massive, and crystallized in about 600
varieties of form, all originating from an obtuse
rhomboid; usually more or less transparent; the
purest varieties refract doubly in a high degree;
lustre between pearly and vitreous, shining; fragments rhomboidal; yields easily to the knife. Sp.
gr. about 2·7.

b. Slate Spar, Schiefer Spar. White, greenish, reddish; pearly; semi-transparent; massive and slaty; curved laminar; soft; greasy. Sp. gr. 2·6.

c. Satin Spar, Fibrous Carbonate of Lime. Composed of fine parallel fibres, sometimes waved; silky, chatoyant; semi-transparent; brittle; harder than common calc. spar. Sp. gr. 2·9.

d. Agaric Mineral, Rock Milk. Whitish; dull; opaque; very friable; stains; often swims a short time on water.

e. Aphrite, Earth-Foam, Silvery Chalk. White; scaly; opaque; friable; sectile; shining; pearly or pseudo-metallic lustre.

f. Stalactitic Carbonate of Lime, Calc Sinter, Stalactite. Occurs mammillated, stalactitic, fungiform, &c., or massive; fine fibrous, diverging, or bladed; pearly or silky; yellowish white.

g. Granular Limestone, Primitive Limestone. Massive, composed of coarse or fine crystalline grains, which are themselves laminar; lustre glimmering; colour extremely various; forms sandy grains during calcination.

h. Common Limestone, Secondary Limestone, Compact Limestone. Massive; compact or granular; fracture large flat conchoidal, and somewhat splintery, rarely earthy; glimmering or dull; colour various; burns to quicklime, without falling to pieces. The three succeeding substances are considered as varieties of Common Limestone:—

Swinestone, Stinkstone. Massive; colour liver brown or black; gives out a strong fœtid odour, when scraped or rubbed; burns into a white quicklime.

Bituminous Limestone. Brown or black; laminar and compact; when rubbed or heated, gives out an unpleasant bituminous odour; by the continuance of heat, it loses both colour and odour, and burns into Lime.

Argillo-Ferruginous Limestone. Occurs massive, and in globular masses; tougher than common limestone; has an argillaceous odour when breathed on, and when burnt is of a buff colour. It is fusible into a slag. Effervesces violently with acids, but is only partially soluble. The *Lias* Limestone, and the *Aberthaw* Limestone, are both argillo-ferruginous varieties.

i. Oolite, Roestone. Occurs in masses composed of small globular distinct concretions, each globule formed of concentric lamellæ; colour greyish brown; opaque; semi-hard; very easily frangible; too impure to burn into Lime. Sp. gr. 2.7.

k. Peastone, Pisolite. Round masses, composed of concentric layers, having each a grain of sand in the centre; white or brown; the concretions vary in size from a pea to a hazle nut, and are sometimes aggregated; opaque.

l. Chalk. Massive; dull; fracture earthy; opaque; soft; light; whitish or greyish; stains the fingers.

m. Marle. Massive; compact or slaty; colour grey; falls to pieces on exposure to the air, and is then plastic with water. Unlike pure carbonate of lime; marle is *easily fusible*, and forms a slag. It effervesces partially in acids, but is only soluble in proportion to its purity.

Bituminous Marle, Bituminous Marle-Slate. Massive; fracture curved or straight slaty; opaque; brownish black; glimmering; soft; meagre.

n. Madreporite, Anthracolite, Prismatic Lucullite. Occurs in large roundish masses, which are composed of prismatic diverging concretions; internal lustre from dull to splendent; greyish black; opaque; soft. Sp. gr. 2.7.—Emits when rubbed a strong smell of sulphuretted hydrogen gas.

o. Tufa, Calctuff. Massive; structure cellular, porous, or spongy; fracture earthy; opaque; dull; light; often incrusting various foreign substances.

551. *Arragonite.*—Agrees in chemical properties, and in double refraction, with carbonate of lime

s

but in some particulars it differs from that substance. Its fracture is coarsely fibrous and imperfectly foliated; cleavage double; fragments not rhomboidal; lustre shining vitreous; yields to the knife, but easily scratches calcareous spar. Sp. gr. 2·9.—Alone, in the matrass, does not change at a low heat, but when red hot, froths and becomes a coarse white porous powder, yielding in the process a very small quantity of water. Arragonite has been divided into three sub-species :—

a. Coralloidal Arragonite, Flos ferri. In snow white branches, consisting of fibrous crystals, diverging from a centre, and sometimes having their points externally apparent.

b. Acicular Arragonite. In six-sided prisms, longitudinally striated, and having two opposite faces larger.

c. Massive Arragonite. With a silky lustre.

552. *Calamine.*—This metallic mineral, see 544, is mentioned here, because it possesses the characteristic of this Order, of being soluble in muriatic acid, and frequently has a great resemblance to an earthy mineral.

553. *White Manganese.*—This is a metallic mineral, see 543, and is placed here merely on account of its possessing the property of effervescing with acids.

554. *Witherite*, Carbonate of Barytes.—Massive, with a divergingly bladed structure, rarely in six-sided prisms, terminated by six-sided pyramids; fracture splintery; brownish white; semi-transparent; glistening resinous; yields easily to the knife. Sp. gr. 4·3.—Decrepitates slightly, and readily fuses into a clear glass, which on cooling becomes a white enamel. Gives a yellow colour to flame.

555. *Strontian*, Strontianite, Carbonate of Strontian.—Occurs rarely in six-sided prisms, terminated by low six-sided pyramids; generally massive; divergingly fibrous structure; shining pearly lustre;

pale green; translucent; yields easily to the knife.
Sp. gr. 3·7.—Fuses with a moderate heat, on the
surface only, assuming a cauliflower appearance and
a dazzling brightness; tinges the blowpipe flame
purple; becomes opaque; the ramified portion acquiring alkaline properties.

556. *Dolomite*, Carbonate of Lime and Magnesia.
—This mineral is not distinguishable by the blowpipe from calcareous spar. It is soluble slowly and
with very little effervescence in cold muriatic acid,
but more rapidly and with considerable effervescence
in hot acid. In this respect it differs from the preceding species of this order. There are the five following subspecies:—

a. Common Dolomite. Massive; in fine concretions; greenish white; translucent; internal lustre
glistening; soft; meagre. Sp. gr. 2·8.

b. Rhomb Spar, Bitter Spar. In middle-sized
rhombs; splendent; grey; fracture foliated; semi-
transparent; scratches calc. spar. Sp. gr. 3·0.

c. Columnar Dolomite. Pale grey; occurs in
pieces about two inches long, interwoven with fibres
of asbestus, and coated with a yellow botryoidal
crust; fragments acicular; longitudinal fracture radiated with delicate cross rents; feebly translucent.
Sp. gr. 2·8.

d. Magnesian Limestone. This is found in great
abundance in England: occurs massive, and differs
from common limestone in having generally a granular sandy structure, a glimmering or glistening
lustre, and a yellow colour.

e. Miemite. Occurs in flat double three-sided pyramids, and in granular and prismatic concretions;
asparagus green; splendent vitreous.

557. *Brown Spar*, Pearl Spar.—Occurs massive,
and in obtuse rhomboids with curvilinear faces;
shining pearly; semi-transparent; pale yellow or
pearl grey; becomes brown on exposure to the air;
yields to the knife, but is harder than calc. spar.

Sp. gr. 2·8.—2·5.—It is infusible; but becomes brown or black before the blowpipe; and, like the varieties of dolomite, is but slowly and quietly soluble in cold muriatic acid. This mineral is frequently confounded with Sparry Iron-ore, which it much resembles in appearance.

ORDER. 2. Fusible before the Blowpipe, Alone.

558. This Order is divided into four Genera, as follows :—

GENUS 1. *Specific Gravity less than* 5·0., *but more than* 3·0.

GENUS 2. *Specific Gravity less than* 3·0., *but more than* 2·5.,

GENUS 3. *Specific Gravity less than* 2·5., *but more than* 2·0.

GENUS 4. *Specific Gravity less than* 2·0.— *Some species supernatant.*

These Genera are all divided into Families, according to the different degrees of hardness of the minerals which they comprehend.

GENUS 1. SPECIFIC GRAVITY LESS THAN 5·0., BUT MORE THAN 3·0.

559. This Genus is divided into five Families :—

FAMILY 1. *Harder than Quartz—Scratch Quartz with ease.*

FAMILY 2. *Rather harder than Quartz—Scratch Quartz with some difficulty—Scratch Felspar with ease.*

FAMILY 3. *Hardness equal to, or exceeding, that*

*of Felspar—Scratched by Quartz—Scratch Win-
dow-glass with ease.*
FAMILY 4. *Softer than Felspar—Scratch Fluor-
spar—Scratch Window-glass feebly.*
FAMILY 5. *Scratched by Fluor-spar.*

—

FAMILY 1. *Harder than Quartz—Scratch
Quartz with ease.*

560. Sp. gr. 4·2.—*Precious Garnet.*—Under the
term Garnet rank many substances (perhaps twelve)
which consist principally of the same elements, and
agree in being found in rhombic dodecahedrons, or
the varieties of that crystalline form; but as these
differ in their specific gravity and other properties,
they are necessarily separated in this synopsis.

Precious Garnet, Almandine, Argillaceous Garnet.—Colour cherry red, often with a tinge of yellow
or blue, and a smoky aspect; translucent or transparent; in rhombic dodecahedrons, with their angles
sometimes replaced; cross fracture conchoidal; lustre shining vitreous.—Before the blowpipe, per se,
becomes brown, but assumes its natural colour on
cooling; fuses without intumescence into a black
glass with a pseudo-metallic surface, and a vitreous lustre on the fracture. When melted with
borax, forms a glass tinged dark by iron. With
mic. salt, intumesces, decomposes, and leaves a skeleton. With soda, intumesces, is decomposed, and
forms a black globule with a pseudo-metallic lustre.
Tried on platinum, exhibits traces of manganese.

561. Sp. gr. 3·8.—*Pyrope,* A variety of the Garnet.—Found in rounded or angular grains, never
crystallized; blood red mixed with yellow; fracture
conchoidal; lustre splendent vitreous; transparent.—
Before the blowpipe, first becomes black and opaque;
if then allowed to cool, it becomes successively dark
green, fine chrome green, colourless, fine fiery red.
Melts difficultly and quietly into a brilliant black

s 2

glass. Yields, when fused with borax, a chrome green glass. Forms, with mic. salt, an opaque green glass, leaving a silica skeleton. Gives a scoria, with soda, without fusing.

562. Sp. gr. 3·0.—*Euclase.*—Crystallized in prisms, longitudinally striated; prevailing colour light green; transparent; cross fracture conchoidal; lustre shining; very frangible; rare.—Unalterable by slight ignition, alone. When heated violently, it becomes opaque, ramifies, and fuses partially into a white enamel. Forms with borax a transparent colourless glass; the fusion being slow, and attended with intumescence and effervescence.

FAMILY 2.   *Rather Harder than Quartz.— Scratch Quartz with some difficulty—Scratch Felspar with ease.*

563. Sp. gr. 3.7.—*Common Garnet.*—Occurs in granular masses, and in roundish crystals, with twelve, twenty-four, or thirty-six sides; colour reddish or blackish brown; semi-transparent; softer than Precious Garnet; glistening. — Before the blowpipe, most of the common garnets fuse with some difficulty, per se, into a blackish or greenish coloured globule. With borax they generally present a glass having the green colour characteristic of iron. The fusion either way, is often accompanied by strong intumescence. On leaf platinum, traces of manganese are produced.

564. Sp. gr. 3·7.—*Melanite*, or Black Garnet, A variety of the common garnet.—Found in lava, in roundish crystals; greyish black; opaque.—Fuses alone into a bright black glass, without intumescence. With borax it fuses slowly into a glass coloured green by iron.

565. Sp. gr. 3·6.—*Cinnamon Stone*, Essonite.—Occurs in splintery fragments, and roundish pieces of an orange brown colour; semi-transparent; contains

many fissures; fracture conchoidal.—Before the blowpipe, it fuses readily into a reddish brown or greenish transparent glass. With mic. salt, decomposes like the other varieties of garnet. With borax fuses with great ease into a transparent glass tinged by iron.

566. Sp. gr. 3·4.—*Grossular*, a rare variety of the Garnet.—Found in varieties of the dodecahedron, with smooth faces; asparagus green; shining; semi-transparent; hard.—Fuses like garnet, and forms a brown vesicular bead.

567. Sp. gr. 3·4.—*Idocrase*, Vesuvian.—Massive, or in groups of crystals; form, a right prism with square base, edges often replaced; fracture small conchoidal; lustre splendent; colour brownish or yellowish green; translucent; refracts doubly; occurs in volcanic rocks.—Alone, before the blowpipe, fuses very readily and with intumescence into a dark coloured glass, which becomes in the exterior flame yellow and transparent. Forms with borax a clear glass coloured by iron. With mic. salt, decomposes and yields a silica skeleton; the glass on cooling becoming opaline.

568. Sp. gr. 3·2.—*Common Schorl*, Black variety of Tourmaline.—Occurs in the same form as the tourmaline, viz. the prism, with many sides and longitudinally striated; also, in acicular, parallel, or diverging concretions; colour velvet black; lustre glistening vitreous; fracture conchoidal; fragile; opaque; pyro-electric with polarity, like the tourmaline. Fuses alone, with strong intumescence, and turns white; afterwards, the intumesced portion melts difficultly into a translucent greyish yellow bead. Effervesces with borax, and melts readily into a clear glass with a tinge of iron. Sharply effervesces and quickly decomposes with mic. salt, leaving a skeleton of silica, and forming an opaline globule. Scarcely dissolves with soda; the product is infusible with additional flux.—A variety of black

tourmaline, or schorl, from Bovey, after intumescing alone, gives a black difficultly fusible scoriaceous mass; but with the fluxes behaves like the above.

569. Sp. gr. 3·2.—*Saussurite*, Variety of Jade.— Massive; leek green; also greyish or whitish green; translucent on the edges, unctuous; extremely tough; fracture splintery; hardness variable.—Alone, before the blowpipe, melts on the angles and edges, but does not fuse entirely.

570. Sp. gr. 3·2.—*Spodumene*, Triphane.—Massive and disseminated, has three distinct cleavages, and a shining pearly lustre; also, occurs fibrous with a silky lustre; translucent; greenish white; brittle.—Gives off water and turns whiter in the matrass. On charcoal, alone, it presents the remarkable appearance of exfoliating into gold coloured scales, and afterwards fuses into a greyish white transparent glass. Intumesces with borax, and forms a transparent tumified mass, but does not melt easily. With solution of cobalt, it affords a glass of a blue colour. The mineral termed *Killinite* is a variety of the preceding.

571. Sp. gr. 3·2.—2·9.—*Tremolite*, Grammatite.— Hardness variable. Divided into three varieties:—

a. Sp. gr. 3·2.—2·9.—Common Tremolite, Crystallized Tremolite. White with a tinge of some other colour; occurs in masses of delicate crystalline fibres which sometimes radiate; also in flat many-sided deeply-striated prisms; translucent; as hard as hornblende; lustre shining pearly; phosphorescent both by heat and friction. Before the blowpipe, in a strong heat, intumesces a little, splits longitudinally, turns white and opaque, and with great difficulty melts, with considerable effervescence, into a distorted and nearly opaque greyish white mass.

b. Sp. gr. 2·8.—Asbestous Tremolite.—Greyish white; in masses of fasciculated groups of minute diverging fibres; fracture fine fibrous, often scopiform; lustre glistening pearly; very soft in mass, but

the concretions are hard; sectile; translucent on the
edges.   When rubbed in the dark, emits a pale red
light; when thrown in powder on hot coals, emits a
green light.—Alone, before the blowpipe, bubbles and
melts with much difficulty into a vitreous mass,
delicately radiated on the surface.   Melts slowly
with borax, but readily with soda, into a transparent
colourless glass.   Fused with solution of cobalt, the
edges are coloured red.

c. Sp. gr. 3·0.—2·8.—Glassy Tremolite, Fibrous
Tremolite.—White with a green, grey, or yellow
tinge; composed of thin prismatic distinct concretions, collected into thick ones; brittle; fracture narrow
radiated with cross rents; translucent; lustre shining pearly; nearly as hard as hornblende; Phosphorescent in a low degree.—Melts before the blowpipe with very great difficulty into a white cellular
scoria.

FAMILY 3. *Hardness equal to, or exceeding,
that of Felspar—Scratched by Quartz—
Scratch Window-Glass with Ease.*

572. Sp. gr. 4·2.—*Gadolinite.*—Scarcely fusible,
see 671.

573. Sp. gr. 3·4.—*Epidote*, Thallite, Pistacite.—
Amorphous and crystallized; yellowish green; crystals often very slender, frequently in groups, and
longitudinally striated; translucent.—Fuses, alone,
at the edges, intumesces, ramifies, and becomes of a
deep brown colour; in a strong flame, it assumes a
round shape, but does not wholly fuse.   With borax,
after intumescing, forms a green bead.   Intumesces
and readily decomposes with mic. salt, leaving a
silica skeleton.—The manganesian variety of the
epidote, which occurs in groups of small prismatic
violet crystals, fuses alone very easily, bubbles and
forms a black glass.   With borax it effervesces and
fuses into a clear glass, which has an amethyst

colour in the outer flame, and a tinge of iron in the
inner; but on cooling becomes colourless.

574. Sp. gr. 3·4.—*Hyperstene*, Labrador Hornblende.—Amorphous, structure lamellar; dark
greenish brown or black, with pseudo-metallic reflections of a copper red; opaque.—Most mineralogists describe this mineral as infusible: Berzelius
gives us the following account:—In the matrass
crackles and gives off water. On charcoal fuses
easily into a greyish green opaque glass. Melts
easily with borax into a greenish glass.

. 575. Sp. gr. 3·3.—*Zoisite*, Variety of Epidote.—
Crystallized in oblique striated four-sided prisms
with incomplete terminations; also amorphous;
structure lamellar; blueish grey; pearly lustre;
translucent.—Before the blowpipe, alone, small bubbles appear, the assay then intumesces and expands,
and its outer edges assume the appearance of transparent yellowish glass; after which the intumesced
mass forms a vitreous scoria of extremely difficult
fusion. Intumesces with borax and forms a clear
glass. With Mic. Salt, intumesces, effervesces, decomposes, and yields a silica skeleton. Melts into a
clear glass with a very little soda; but with a large
portion forms a white infusible mass. Gives traces
of manganese on platinum foil.

. 576. Sp. gr. 3·2.—*Axinite*, Thumerstone.—In flat
crystals, with remarkably sharp edges, sometimes
pyro-electric; colour violet; transparent; splendent;
fragile.—Fuses into a green glass with intumescence.
The exterior flame renders this glass black. With
borax, fuses into a glass tinged by iron, but having
an amethystine hue in the outer flame. Becomes
green with soda, then fuses into a blackish glass,
with a semi-metallic lustre.

577. Sp. gr. 3·2.—*Bronzite.*—See 677.

578. Sp. gr. 3·2.—*Basaltic Hornblende.*—Occurs
in six-sided prisms, terminated by two three-sided
pyramids; deep black; opaque; splendent vitreous;

often magnetic; structure laminar. Fuses into a
shining black glass.

579. Sp. gr. 3·2.—*Haüyne*, Latialite.—Massive,
and in minute rhombic dodecahedrons; blue and
blueish green; shining vitreous; very brittle; fracture conchoidal.—On charcoal, loses its colour, and
fuses into a blebby glass; with borax effervesces, and
fuses into a transparent glass, which on cooling becomes yellow, and if saturated, opaque.

580. Sp. gr. 3·2.—2·9.—*Tremolite.*—See 571.—
Some of the varieties agree in hardness with the
minerals of this family.

581. Sp. gr. 3·0. — *Jade*, Nephrite.—Hardness
variable; massive; very tough; splintery fracture;
leek green and greenish white; lustre and feel
greasy; translucent on the edges.—In the matrass
gives off water, and blackens. On charcoal, fuses
with difficulty into a white glass. Fuses with borax,
into a transparent glass; with soda, into a semi-
transparent mass. Partially fuses with solution of
cobalt, forming a black glass.—*Axe-stone* is a variety
of Jade, which differs from the above, in being of a
darker green colour, not so transparent, nor so tough;
and in having a slaty structure.

582. Sp. gr. 3·0.—*Amblygonite.*—Occurs in rhombic prisms; green; semi-transparent; reducible by
cleavage to thin laminæ. In the matrass, gives off
water, and fluoric acid. Fuses easily on charcoal,
into a transparent glass, which becomes opaque on
cooling. With boracic acid and iron, produces phosphuret of iron.

FAMILY 4. *Softer than Felspar — Scratch
Fluor-spar—Scratch Window-Glass feebly.*

583. Sp. gr. 3·8.—*Yenite*, Lievrite.—Occurs amorphous and in rhombic prisms, longitudinally striated; blackish brown; opaque; shining semi-metallic.—Gives off water in the matrass. Fuses on
charcoal into a black glass, which is rendered mag-

netic by the interior flame, provided the assay be
not heated to redness. Fuses readily with borax
into a glass dark in colour, and nearly opaque.
Forms a black glass with soda, and shows traces of
manganese on leaf platinum.

584. Sp. gr. 3·6.—*Scapolite*, Pyramidal Felspar,
Paranthine.—This mineral is very liable to decomposition; so that its specific gravity and hardness
vary much. Great confusion exists in the various
descriptions of its subspecies, of which, according to
Jameson, there are four:—

a. Radiated Scapolite, Bergmannite.—Grey or
greenish white; in distinct concretions; fibrous, radiated, and in rectangular striated prisms aggregated; shining; translucent. Sp. gr. 2·9.—2·5.—Before the blowpipe, per se, becomes white and opaque,
and then fuses into a blebby colourless glass. With
borax and mic. salt, effervesces considerably, and
fuses into a transparent glass. Forms a transparent
glass with soda, which continues fusible when more
flux is added.

b. Foliated Scapolite, Wernerite, Gabronite.—
Grey, green, and black; massive and in octohedral
prisms; splendent vitreous; translucent; brittle.

c. Compact Red Scapolite.—Darkish red; in acicular crystals, often curved; rough; dull; nearly
opaque.

d. Elaolite, Fetstein, Natrolite.—Dark blue, grey,
or brownish red; granular; massive; translucent;
shining resinous, sometimes opalescent. Sp. gr. 2·6.
Gelatinizes in acids when powdered.—Gives off a
little water in the matrass, and melts easily, with
effervescence, on charcoal, into a blebby colourless
glass.

e. Dipyre.—Berzelius says, this appears from its
pyrognostic habitudes to be a true scapolite; notwithstanding Vauquelin's analysis.

585. Sp. gr. 3·3.—*Sahlite*, Laminar Pyroxene,
Malacolite, Variety of Augite.—Amorphous and in

prismatic crystals.; structure distinctly laminar ;
cleavage three-fold; greenish grey; often translucent; splendent.—Alone, it intumesces, and melts
into a transparent colourless glass.. With borax;
fuses easily into a transparent glass.. Intumesces
with a small portion of soda, and fuses into a clear
glass. If more soda is added in small quantities,
at each addition, fusion is preceded by intumescence, and the assay becomes opaque and less fusible.

586. Sp. gr. 3·3.—*Augite*, Pyroxene.—Prevailing
colour dark green; crystallized, amorphous, and in
grains.; often nearly opaque; found in volcanic and
basaltic rocks. Before the blowpipe, melts into a
black enamel with extreme difficulty, and only in
minute portions; this property distinguishes it from
hornblende,. which it much resembles in appearance.

b. Diopside, Mussite, Alalite, Variety of Augite
passing into Sahlite.—In crystals; prismatic; green;
translucent; shining; longitudinally striated; Sometimes in groups,. diverging or straight.—Behaves
much like Sahlite before the browpipe.

587. Sp. gr. 3·3.—*Common Hornblende*, Amphibole.—Amorphous, sometimes in flat six-sided prismatic crystals ; colour greenish black ; fracture crystalline,. exhibiting fibres confusedly aggregated ;
opaque when black ; translucent on the edges when
green ;. lustre shining pearly ; powder green ; exhales
a bitter smell when moistened.—Before the blowpipe,
melts with great ease into a blackish coloured glass.

588. Sp. gr. 3·3.—*Common Actynolite*, Actinote;
considerated a pure variety of Amphibole or Hornblende.—Leek green ; in wedge-shaped concretions,
composed of acicular prisms ; fracture fibrous and
scopiform (often very finely so); singularly harsh to
the touch ; translucent ; shining silky ;—Melts into
a, greenish grey glass.

b. Glassy Actynolite is the variety that occurs in
acicular crystals, with a grass green colour, and a

T

shining vitreous lustre.—When the crystals are extremely minute, it passes into

   c. Asbestous Actynolite, which approaches very nearly to the state of amianthus. It is softer than the other varieties, of a greenish brown colour, and the fibres are particularly distinct. It melts before the blowpipe into a black or grey scoria.

   d. Actynolite sometimes occurs granular.

   589. Sp. gr. 3·1—*Hornblende Slate.*—Occurs in beds, in clay state, and is very common; colour from greenish to raven black; massive; internal lustre shining pearly.—Fracture, in the great slaty, in the small promiscuous radiated.

   590. Sp. gr. 3·1.—*Smaragdite,* Bright Green Diallage.—Bright emerald green; lustre glistening silky; feebly translucent; massive and disseminated; structure fibrous laminar; scarcely so hard as glass.—Fusible into a grey or greenish enamel.

   591. Sp. gr. 3·0.—*Basalt.*—Occurs amorphous, columnar, in thick concentric lamellar concretions, vesicular, and amygdaloidal; greyish and blueish black; texture fine grained, composed of minute grains or crystals; the compact varieties pass into clinkstone, the granular into greenstone; fracture somewhat conchoidal; dull and opaque; breaks with difficulty, but yields to the knife.—Easily melts before the blowpipe into a black glass.

FAMILY 5.   *Scratched by Fluor-spar.*

   592. Sp. gr. 4·5.—*Sulphate of Barytes,* Heavy spar, Baroselenite, Ponderous Spar. —Occurs in veins, massive and crystallized; structure lamellar, with joints parallel to the faces of a right rhomboidal prism; prevailing colour greyish or yellowish white, sometime inclining to red or blue; from transparent to opaque; lustre shining pearly or vitreous; fragile; harder than carbonate of lime, but less hard than fluor-spar; refracts doubly in a particular direction.—Soluble in no acid but the sulphuric, and

precipitated from that by water. Before the blow-pipe, decrepitates with violence, and by the conti-nuance of a strong heat, melts into a hard white enamel. In the interior blue flame, it fuses and de-composes, forming sulphuret of barytes, which ex-hales a slight hepatic odour when moistened, or when laid on the tongue gives the flavour of sulphuretted hydrogen gas or of putrid eggs.

a. Crystallized Sulphate of Barytes, Lamellar Heavy Spar.—The most common forms are, the oc-tohedron with cunieform summits, the four or six-sided prism, the hexangular table with bevelled edges; often large; transparent; shining vitreous.

b. Columnar Heavy Spar, Stick Spar.—Rhombic prisms, aggregated into columns; greenish white; translucent; shining pearly; structure lamellar.

c. Fibrous Heavy Spar.—Botryoidal or reniform masses; structure fibrous; shining resinous; ches-nut brown; nearly opaque; brittle.

d. Radiated Sulphate of Barytes, Bolognian Stone. —Globular masses, rough externally; composed of minute fibrous crystals diverging from a centre; in-ternally grey; shining; translucent on the edges; soft; fragments wedge-shaped; remarkably phos-phorescent when heated.

e. Granular Heavy Spar.—Occurs in masses com-posed of small grains, which have a lamellar struc-ture; glistening; blueish white.

f. Cawk.—Massive; fracture coarse earthy; opaque; nearly dull; white with a tinge of yellow or red; soft; brittle. Sp. gr. 4·8.—Often containing small veins of galena.

g. Earthy Sulphate of Barytes.—Dull or glimmer-ing particles, which cohere very slightly, and are meagre to the touch and heavy.

h. Fœtid Sulphate of Barytes, Hepatite.—Occurs in globular masses which have a lamellar or radiated structure; yellow, brown, or black; yields a fœtid odour when rubbed or heated.

593. Sp. gr. 4·3.—*Carbonate of Barytes*, Withe-rite. Soluble in acids. See 554.

594. Sp. gr. 4·0.—3·6.—*Sulphate of Strontian*, Ce-lestine.—Occurs massive and crystallised; primitive form a right rhomboidal prism; structure lamellar; prevailing colour white, inclining to yellow, red, or delicate blue; translucent or transparent; shining pearly or resinous lustre; refracts doubly; harder than heavy spar; scratches calcareous spar; softer than fluor-spar.—Its most remarkable character is the rose-red colour which it communicates to flame—thus, if a little spirits of wine be inflamed and pul-verised strontian ore (either sulphate or carbonate) be added by degrees, a fine red flame will be pro-duced. By a similar process barytes yields a yellow flame. Before the blowpipe, exposed to the outer flame, it decrepitates and melts into a milky white enamel, which, in the inner flame, spreads about, decomposes, becomes infusible, and produces a hepa-tic mass. When this mass is cold, it exhales the odour of rotten eggs, and has a peculiar hepatic and acrid taste. On platinum foil, with muriatic acid, it partially dissolves, producing a solution which on evaporation to dryness yields a salt. If this salt is laid on a piece of paper wetted with alcohol and set on fire, the flame near the salt will be tinged red. Effervesces with borax, and melts into a transparent glass which becomes brown or yellow on cooling.

595. Sp. gr. 3·7.—*Carbonate of Strontian.*—Solu-ble in acids. See 555.

596. Sp. gr. 3·2.—*Apatite*, Phosphate of Lime.—There are three varieties of this mineral:—

a. Foliated Apatite, Phosphorite of Kirwan, Crys-tallized Phosphate of Lime. — Prevailing colour white, passing to yellow, red, green, or blue; often several colours occur in the same specimen. Some-times massive, and in lamellar distinct concretions; generally crystallised in six-sided prisms which are short, truncated, striated, translucent, and splendent

resinous; electric by heat and friction; softer than
fluor spar, but harder than calcareous spar. When
thrown on glowing coals, emits a pale green phosphoric light. Dissolves very slowly and without
effervescence in nitric acid.

  b. Conchoidal Apatite, Asparagus Stone, Morox-
ite. — Prevailing colours green and blue; occurs
massive and in granular distinct concretions; but
generally in long six-sided prisms with pyramids;
splendent vitreous externally; shining resinous internally; fracture conchoidal; sometimes does not
phosphoresce with heat.—Before the blowpipe, a solid fragment remains unaltered, but a spangle, in a
very strong heat, fuses on the edges into a colourless transparent glass. Apatite is one of the most
difficultly fusible minerals; indeed, mineralogists
generally, but erroneously, say that it is infusible.
Melts slowly with borax into a clear glass, which is
rendered milk white by flaming; or, if the flux is
saturated with the assay, becomes opaque on cooling.
Fuses in large quantity with mic. salt, and forms a
clear glass which crystallizes on cooling. Effervesces
and boils with soda; the flux then sinks into the
charcoal, and leaves a white mass.

  c. Phosphorite.—Divided into two kinds:—1.
Common Phosphorite, Massive Apatite. Colour greyish and yellowish white; massive and in distinct
concretions which are thin curved lamellar; surface
uneven and drusy; dull or glistening; fracture uneven; cleavage floriform; opaque. When rubbed
in an iron mortar, or pulverized and thrown on
glowing coals, it emits a green-coloured phosphoric
light.—In the matrass, yields a little water; afterwards behaves before the blowpipe like the preceding; but it is more easily fusible per se, forming a
white enamel.—2. *Earthy Phosphorite.* Greenish
grey; consists of dull dusty particles, which cohere
but loosely, soil slightly, and feel meagre. It phosphoresces when laid on glowing coals.

597. Sp. gr. 3·2.—*Fluor-spar*, Fluate of Lime.—Occurs in metallic veins, amorphous and crystallized; colours violet blue, emerald green, yellow, and various; frequently clear and bright; of different degrees of transparency and lustre; considerably harder than calcareous spar, but not hard enough to scratch glass; when placed on a hot iron, it emits a blue or greenish light.—Alone, in the matrass, gently heated, exhibits a greenish light in the dark; heated powerfully, it decrepitates violently and yields some water. On charcoal, in a good heat, it fuses into an opaque globule. Melts with borax and mic. salt into a clear glass which becomes opaque when saturated to a certain extent. With soda, fuses into a transparent glass which subsequently becomes an enamel :—

a. Crystallized Fluor. Occurs in aggregated cubes generally modified; the primitive form, the octohedron, can be obtained by cleavage; general colours purple and green; splendent and transparent; often very beautiful.—Chlorophane is a variety of crystallized fluor.

b. Massive Fluor, Blue John. Nodulous and kidney-shaped; structure crystalline fibrous diverging; colours purple and yellowish white in concentric zones.

c. Compact Fluor. Has a flat conchoidal or splintery fracture; is harder than common fluor, and translucent; but the lustre and colours are faint.

d. Earthy Fluor. Occurs pulverulent and encrusting.

GENUS 2. SPECIFIC GRAVITY LESS THAN 3·0., BUT MORE THAN 2·5.

598. This Genus is divided into four Families:—

FAMILY 1. *Hardness rather exceeding that of Quartz, or nearly the same—Not scratched by Quartz—Scratch Felspar.*

FAMILY 2. *Hardness about equal to that of Felspar*
   —Scratched by Quartz—Scratch Window-glass
   with ease.

FAMILY 3. *Softer than Felspar—Harder than*
   Fluor-spar.

FAMILY 4. *Scratched by Fluor-spar.*

—

FAMILY 1.   *Hardness rather exceeding that*
   of Quartz, or nearly the same—Not scratch-
   ed by Quartz—Scratch Fluor-spar.

599. Sp. gr. 2·7.—*Emerald.*—This beautiful mineral is characterized by its pure vivid green colour,
commonly called emerald-green ; occurs crystallized
in low six-sided prisms, sometimes truncated on the
edges or angles ; cross fracture conchoidal ; lustre
vitreous splendent ; refracts doubly ; transparent or
translucent.—Alone, before the blowpipe, a gentle
heat produces no change. If a strong heat is applied for a considerable length of time to a thin
scale, it becomes rounded on the edges, and forms a
whitish frothy glass. The transparent varieties become milk white when the heat is concentrated.
With borax it melts into a diaphanous glass, sometimes free from colour, sometimes having a fine pale
pleasant green tinge. With mic. salt, fuses slowly
but perfectly, leaving no skeleton of silica. The
chrome-green emerald affords a green glass. Fuses
with soda into a transparent colourless glass. The
yellowish emerald with a granular fracture, gives, in
the reducing experiment, traces of tin. Produces
with the solution of cobalt, an impure slightly blueish colour.

600. Sp. gr. 2·6.—*Beryl,* Aquamarine, a Variety
of Emerald.—The colour of the beryl is paler than
that of the emerald, and the prisms are longer and
frequently larger ; it is also harder, and more readily
yields to cleavage.—Before the blowpipe, it behaves
like the emerald.

601. Sp. gr. 2·6.—*Iolite*, Dichroite, Sapphir d'Eau, Pelium, Peliome, Steinheilite.—Colour deep violet, but when viewed by transmitted light at right angles to the axis of the prism, yellowish brown; occurs amorphous and in six-sided prisms; fracture conchoidal; shining vitreous lustre; translucent to opaque; sometimes doubly refractive, and electric by rubbing.—Before the blowpipe, alone, at a low heat, no change occurs; in a strong heat, slowly melts on the edges into a glass which is not frothy; the assay retaining its original colour and degree of transparency. Fuses into a clear glass with borax. With soda no solution is effected; a small portion of this flux produces a dark grey glassy scoria.

602. Sp. gr. 2·7.—*Adularia*, Moonstone, a variety of Felspar.—Prevailing colour greenish or greyish white; splendent pearly lustre; highly translucent; is the hardest and purest variety of felspar; occurs massive and crystallized in prisms; structure distinctly laminar; yields to mechanical cleavage; when cut in a rounded form and polished, presents beautiful chatoyant pearly reflections.—In the matrass, felspar yields no water if transparent; but a large quantity is given out from the interstices of cracked opaque felspar. On charcoal, alone, in a bright heat, it turns glassy, whitish, and semi-transparent; and with great difficulty melts on the edges into a translucent frothy glass or enamel. Fuses with borax into a clear glass, but slowly and quietly. If pulverised and mixed with mic. salt, it yields a globule which becomes opaline on cooling, leaving a skeleton of silica. Effervesces with soda, and slowly melts, giving a glass, which is transparent, but rarely free from blebs. Gains a blue colour from solution of cobalt, but merely upon the fused edges.

603. Sp. gr. 2·5.—*Colophonite*, a variety of Garnet. —Yellowish, reddish, and blackish brown; lustre resinous; occurs amorphous, and in four-sided prisms; translucent; very easily frangible.—Before

the blowpipe, fuses very easily into a black-coloured bead. With borax it melts into a yellowish glass.

FAMILY 2. *Hardness about equal to that of Felspar — Scratched by Quartz — Scratch Window-glass with ease.*

604. Sp. gr. 2·9—2·6.—*Prehnite,* one of the Zeo-lites.—Prevailing colour pale green ; occurs in botry-oidal or globular concretions ; fracture finely radi-ated fibrous ; found also in low six-sided pyramids, or tables, often in radiated groups, lustre pearly shining, more or less transparent. Becomes electric by heat. Very plentiful in the trap rocks near Glasgow. A variety, in small transparent rhombic tables, is called Koupholite.—Gives off a little water in the matrass, but preserves its transparency until, by exposure to a high temperature, it intumesces and begins to fuse. Specimens of Koupholite, which, by exposure to the air in mineralogical collections, have got their pores filled by dust, become blackish and give out an empyreumatic odour. The crystal-line laminæ, however, recover their transparency and melt into a blebby white glass. Fuses with borax into a clear glass, which, if saturated, becomes turbid and nearly infusible. With soda forms a semi-vitreous scoria.

605. Sp. gr. 2·6.—*Felspar.*—One of the most abundant of simple minerals, being the principal constituent of granite. Occurs massive and crystal-lized in prisms. Structure lamellar, with joints in two directions, at right angles to each other.—For its pyrognostic habitudes, see Adularia (602).

a. Common Felspar. Crystallized, massive, dis-seminated ; lustre on the natural joints shining, be-tween vitreous and pearly ; cross fracture uneven, glimmering ; prevailing colours white and red ; more or less translucent.—*Albite* or *Cleavelandite* is a vari-ety of Common Felspar. Its colours are greyish white or dingy red, and its structure broad promiscuous fibrous.

b. Labrador Felspar, Opalescent Felspar. Found in rolled masses, never crystallized; structure curved lamellar; smoke grey colour, but exhibits, by reflection, a rich play of yellow, blue, and red colours, with a splendent semi-metallic lustre; translucent.

c. Glassy Felspar. In embedded crystals, which appear as if cracked in many places; semi-transparent; greyish white; lustre splendent vitreous; softer than common felspar.

606. Sp. gr. 2·5.—*Petalite.*—Nearly resembles quartz; white with a tinge of purple; translucent; coarsely fibrous structure; a rare mineral.—Behaves before the blowpipe like felspar.

607. Sp. gr. 2·5.—*Porcelain Jasper*, Porcellanite. —Occurs massive; structure slaty; presents various shades of blue, grey, red, and yellow, clouded and spotted; fracture smooth glistening; resembles a semi-vitrified brick; opaque.—Before the blowpipe, melts into a spongy yellowish-white semi-transparent mass. All the other varieties of jasper are infusible.

FAMILY 3. *Softer than Felspar—Harder than Fluor-spar.*

608. Sp. gr. 3·0.—*Datholite*, Borate of Lime, Botryolite.—Greyish or greenish white; translucent; massive, and in rhombic prisms truncated on the edges and angles; lustre shining vitreous. Botryolite occurs in botryoidal concretions, which have a lamellar concentric structure, and are composed of delicate radiated fibres. Gelatinizes with acids. Becomes soft, white, and opaque in the flame of a candle.—In the matrass, gives off a little water. Before the blowpipe on charcoal, intumesces like borax, and melts into a clear glass, with a pale rose or green colour. Acts in a similar manner with borax. Gives, with solution of cobalt, an opaque blue glass.—*Expt.* Moisten the powder of this mine-

ral with muriatic acid, and let it dry on a slip of
thin paper; then wet the paper with alcohol and
set it on fire,—the flame, towards the end of the
combustion, will be tinged green.

609. Sp. gr. 2·9.—2·6.—*Wakke*. Wacké.—Massive,
either solid, cellular, or amygdaloidal; greenish grey;
dull; earthy; opaque; greasy to the touch;. gives
out an argillaceous odour when breathed upon.—
Before the blowpipe, melts into a greenish porous
slags.

b. Iron Clay. This may be considered a variety
of the above. Colour reddish brown; usually cellular or amygdaloidal; easily frangible. Agrees
with Wakke in other respects.

610. Sp. gr. 2·9.—*Tabular Spar*.—Massive and in
prismatic concretions; white; shining pearly; often
friable; phosphorescent when scratched by a knife.
Effervesces quickly when dropped in nitric acid, and
then falls into powder.—Alone, before the blowpipe
on charcoal, melts on the edge into a semi-transparent colourless glass. In a very strong heat, complete fusion takes place.

611. Sp. gr. 2·9.—2·5.—*Boracite*, Borate of Magnesia.—Occurs in solitary crystals; form, the cube and
its varieties; greyish or greenish white; translucent;
splendent vitreous; pyro-electric, the solid angles
diagonally opposite being one positive the other negative.—Before the blowpipe, alone, on charcoal
intumesces and fuses, the globule is yellowish and
transparent while hot, but appears white and opaque
on cooling, from its surface being bristled over with
needle-formed crystals. Forms with borax a transparent glass tinged by iron. With a small quantity
of soda fusion ensues, and the assay, when cooling,
forms crystals with broad facets as perfect as those
of phosphate of lead. If the *experiment* with Datholite, (described above,) be repeated with Boracite,
after it has been decomposed by soda on charcoal,
a similar green flame will be produced.

612. Sp. gr. 2·9.—*Arragonite.*—Soluble in Muriatic Acid, see 551.

613. Sp. gr. 2·8.—*Lepidolite*, Scale-stone.—Consists of numerous small scales, or hexagonal plates, aggregated into a mass; lilac purple or pearl grey; lustre shining; translucent; unctuous; may be cut with a knife.—In the matrass, alone, gives off water loaded with fluoric acid. Intumesces on charcoal, and easily melts into a transparent colourless blebby globule. Fuses readily and largely into a transparent glass with borax. Produces a blue colour when fused with solution of cobalt.

614. Sp. gr. 2·8.—*Lapis Lazuli*, Lazulite, Ultramarine Stone.—Colour bright azure blue, frequently with white spots and veins of yellow pyrites; generally massive; nearly opaque; possesses little lustre; has a fine grained fracture: effervesces a little with acids, and, if previously calcined, gelatinizes.—Before the blowpipe it melts into a blackish mass, which, at a high temperature, forms a white enamel.

b. Azurite, False Lazulite.—Indigo blue; massive, in oblique prisms; laminar; opaque.—Very difficultly and but partially fusible, alone, into a whitish glass with blue and green spots; a mineral of little importance.

615. Sp. gr. 2·7.—*Novaculite*, Whetslate, Hone, Turkey-stone.—Greenish or yellowish grey; massive; slaty; dull; greasy; is a compact variety of clay-slate, used for sharpening steel instruments.—Before the blowpipe it becomes white, and acquires a vitreous glazing.

616. Sp. gr. 2·7.—*Chabasie*, Chabasite, one of the Zeolites.—Occurs crystallized in obtuse rhomboids, in the cavities of basaltic amygdaloid; white or greyish, pale red superficially; highly translucent; splendent vitreous; scratches glass feebly.—Alone, before the blowpipe, it melts, with little ebullition, into a white spongy enamel.

617. Sp. gr. 2·6.—*Clinkstone.*—Occurs massive,

columnar, or with a thick slaty structure; gives a
ringing metallic sound when struck with a hammer;
dark greenish grey; glimmering; nearly opaque;
fracture splintery.—Before the blowpipe, fuses easily
into a glass nearly colourless.

618. Sp. gr. 2·6.—*Elaolite.*—A variety of Scapolite, of low specific gravity. See 584, *d.*

619. Sp. gr. 2·6.—*Meionite.*—Occurs in groups of
small crystals; form, a four-sided prism terminated
by tetrahedral pyramids, of which the edges are
generally truncated; structure laminar, with a rectangular cleavage; white; transparent; shining
vitreous; smooth.—Alone, before the blowpipe, a
thin splinter foams at its extremities, after which
the whole mass bubbles considerably and for a long
time, producing in the end a colourless blebby glass.
Melts slowly with borax, with considerable effervescence, into a clear glass. Intumesces with soda,
of which a large dose is requisite, and forms a glass
which is opaque on one side, but in the end becomes
transparent.

FAMILY 4. *Scratched by Fluor-spar.*

620. Sp. gr. 2·9.—*Cryolite.*—Occurs in lamellar
masses, colour greyish white or brown; glistening
vitreous; translucent; transparent when immersed
in water; melts in the flame of a candle; rare.—
Before the blowpipe it at first runs into a very liquid
fusion, then hardens, and at length assumes the
appearance of a slag.

621. Sp. gr. 2·9.—*Anhydrite,* Anhydrous Gypsum.
—Of this mineral there are several varieties; the
pyrognostic characters of which are as follows:—
Yield no moisture in the matrass; melt difficultly,
in the oxidating flame, into a white enamel, when
held in the forceps. On charcoal, decompose in a
good reducing flame; after which they act on brazil-
wood paper as alkalies, and when moistened, exhale
the odour of liver of sulphur. Effervesce with bo-

rax, and melt into a clear glass, having a brownish yellow colour when cold. With glass of soda, develope the odour of liver of sulphur :—

a. Muriacite. Occurs crystallized in rectangular prisms; structure laminar; yields to mechanical cleavage; lustre shining pearly; colours white, violet, blueish; transparent or translucent; doubly refractive; scratches calcareous spar, but is scratched by fluor-spar.

b. Granular Anhydrous Gypsum, Scaly Anhydrite. Occurs in massive concretions, with confusedly foliated structure; greyish white; translucent on the edges.

c. Fibrous Anhydrous Gypsum. Massive; structure fibrous and sometimes radiated; colours greenish grey, blueish, or reddish; translucent on the edges.

d. Compact Anhydrous Gypsum. Massive, contorted and reniform; translucent; fracture splintery or flat conchoidal; harder than the preceding varieties.

e. Siliciferous Anhydrous Gypsum, Vulpinite. Occurs in distinct concretions, structure laminated; translucent on the edges; has more lustre than the preceding; soft; brittle; greyish white veined with blue.

622. Sp. gr. 2·9.—*Pinite*.—Occurs in six-sided prisms; blackish grey or green, surface brown; opaque; glistening; sectile. Alone, on charcoal, turns white and melts on the edges into a white blebby glass; some varieties become covered with coloured spots; and others easily fuse into a black glass. Pinite is a scarce and unimportant mineral.

623. Sp. gr. 2·9.—*Karpholite*, Carpholite.—Occurs in minute fibrous crystals radiated; yellow; translucent; shining; pearly; brittle.—In the matrass, yields water and fluoric acid. Alone, on charcoal, before the blowpipe, intumesces, whitens, and slowly fuses into a brown opaque glass; with borax fuses into a transparent glass, which possesses

the colour of manganese in the exterior flame, but turns green in the interior.

624. Sp. gr. 2·8.—*Polyhallite.*—Amorphous; compact, colourless, and transparent; or curved lamellar fibrous, with a brick red colour and translucent; brittle.—In the flame of a common candle it immediately becomes an opaque brown mass. In the matrass, gives off water and loses its colour. On charcoal melts into an opaque reddish globule, which in the inner flame, whitens, congeals, and leaves an empty shell. The globule has a saline and hepatic taste.

625. Sp. gr. 2·8.—2·6.—*Chlorite*, Talc Chlorite.—Green is the prevailing colour, varying from dark green to light greyish green, and generally dull; occurs crystallized, amorphous, and in microscopic hexagonal scales or prisms; opaque; lustre shining pearly; soft; unctuous.—In the matrass, gives off water, and (when the glass begins to fuse) fluoric acid, known by its turning brazil-wood paper yellow, and depositing silica on the glass. On charcoal, alone, melts into a black globule with a dull surface. Gives a dark green glass with borax. Decomposes with mic. salt, forming a glass tinged by iron, and leaving a silica skeleton. Does not fuse with soda, nor intumesce, yet the edges of the assay become rounded.

a. Earthy Chlorite. Occurs in friable or loosely adhering grains; imbedded or encrusting.

b. Common Chlorite. Essentially the same as *a*, but the grains cohere more firmly; glistening.

c. Chlorite Slate. Occurs in beds; structure slaty; glistening; resinous.

d. Foliated Chlorite. Occurs crystallized in flat six-sided prisms, which are readily divisible into hexagonal curved laminæ; lustre shining resinous; translucent on the edges; the crystals often grouped in conical masses.

e. White Silvery Chlorite. Pearly greenish-white

scales, which occur in masses, but adhere so slightly
that they may be separated with the finger.

f. Green Earth. Found in globular masses or
lining the cavities of amygdaloid; fracture earthy;
colour lively blueish green; dull; streak glistening;
soft; unctuous; light.

626. Sp. gr. 2·7.—*Mica.*—This mineral is characterized by the facility with which it divides into
extremely thin smooth shining plates or laminæ,
which are transparent, flexible, and highly elastic;
occurs in masses, rarely in tabular crystals, and
abundantly disseminated in grains; colours various;
lustre shining, often pseudo-metallic; scratched by
the knife, but the edges scratch glass; smooth but
not unctuous.—Mica from different localities, differs
considerably in its pyrognostic characters; but it may
be said, in general, to be difficultly fusible before
the blowpipe into a grey enamel.

627. Sp. gr. 2·7.—*Soapstone.*—Colour grey, mottled
with dull purple; massive; commonly described as
a variety of steatite (712), which it much resembles,
but it is much softer; indeed, when first raised it
may be kneaded like dough, but it becomes harder
on exposure to air; it cracks and falls to pieces in
hot water.—Before the blowpipe, it fuses into a
white and somewhat translucent enamel.

628. Sp. gr. 2·6.—*Clay-Slate,* Argillaceous Schis-
tus.—Occurs massive; structure slaty; fracture compact; prevailing colours blueish and greenish grey;
glimmering or glistening; generally opaque; does not
adhere to the tongue.—Fuses before the blowpipe
into a black slag. Roofing slates, slates for writing
on, and pencils for writing with, are varieties of this
mineral.

629. Sp. gr. 2·6.—*Shale,* Slate-Clay.—Massive;
slaty; grey; dull; opaque; meagre; brittle; disintegrates on exposure to the air, and by degrees becomes plastic.—Fusible into a slag.

b. Rottenstone appears to be decomposing shale;

it is dirty grey ; dull ; earthy ; soft ; meagre ; and
fœtid when rubbed or scraped.

630. Sp. gr. 2·5.—*Serpentine.*—Although this is
usually classed as a simple mineral, the great variation in the composition of the rocks to which the
term is applied, shows that it is an indefinite earthy
compound. There are two varieties :—

a. Noble Serpentine, Precious Serpentine. Occurs
massive; fracture splintery, passing into conchoidal;
colour yellowish green, uniform throughout; translucent; glistening; unctuous; soft; sectile.—In the
matrass, gives off water and turns black. On charcoal, turns white, and in a good heat, melts on the
thin edges into an enamel. Slowly melts with borax
into a clear glass coloured green. Gives a red colour
with solution of cobalt.

b. Common Serpentine, Serpentine Rock. Differs from the preceding in being less pure ; colours
green, red, and various, intermixed in stripes,
dots, &c.; generally opaque and dull; structure
compact ; hardness extremely variable ; sometimes
magnetic.—Before the blowpipe, it behaves like the
preceding.

GENUS 3. SPECIFIC GRAVITY LESS THAN 2·5.,
BUT MORE THAN 2·0.

631. This Genus is divided into two Families :—

FAMILY 1. *Scratched by Felspar.—Scratch Window-glass and Fluor-spar.*

FAMILY 2. *Softer than Fluor-spar.*

FAMILY 1. *Scratched by Felspar—Scratch
Window-glass and Fluor-spar.*

632. Sp. gr. 2·5. — *Apophyllite,* Fish-eye-stone,
Ichthyophthalmite.—Occurs in square prisms of various heights; when tabular, they are often cellular

aggregated; the surfaces produced by cleavage are splendent, pearly, iridescent; fracture conchoidal and glistening vitreous; reddish or greenish white; transparent; very fragile; hardness variable. Feebly electric by friction. When pulverised, gelatinizes in nitric acid.—In the matrass, yields much water and whitens. On charcoal, immediately exfoliates, swells up like borax, and melts, with continued intumescence, into a blebby colourless glass. Melts easily with borax, into a diaphanous glass, which becomes opaque by flaming.

633. Sp. gr. 2·4.—*Obsidian*, Volcanic Glass.—Occurs massive or in roundish pieces; colour dark greenish or brownish black; sometimes semi-transparent, but generally translucent on the edges; lustre splendent vitreous; fracture large and perfectly conchoidal; very brittle; breaks into very sharp edged fragments; bears a striking resemblance to dark coloured glass.—Before the blowpipe, melts into a greyish or greenish vesicular glass.

634. Sp. gr. 2·3.—*Harmotome*, Cross Stone, one of the Zeolites.—Occurs crystallized; primitive form, a pyramidal octohedron; secondary form, small four-sided prisms, terminated by four-sided pyramids, which are often connected two and two in the shape of a cross; colour greyish white; semi-transparent; lustre shining pearly; scratches glass.—In the matrass, gives off water and becomes opaque. On charcoal, alone, fuses easily and without intumescence, into a clear glass. Forms a colourless glass with borax, but with much difficulty.

635. Sp. gr. 2·3.—*Pearl-stone*, Pearly Obsidian.—Occurs in masses of globular concretions, which are composed of thin concentric laminæ; surface of the concretions smooth, shining, pearly; colour grey; translucent on the edges; almost friable; scarcely scratches glass.—Before the blowpipe, it first swells, splits, and becomes white, then, with some difficulty, melts into a whitish frothy glass.

636. Sp. gr. 2·3.—*Pitchstone*, Retinite.—This substance bears a striking resemblance to pitch ; occurs massive and in prismatic and curved lamellar concretions ; structure sometimes slaty ; fracture imperfectly conchoidal ; colours various shades of grey, green, blue and black, muddy ; lustre shining, resinous or pitchy ; translucent on the edges.—Fusible before the blowpipe, into a grey frothy enamel.

637. Sp. gr. 2·3.—*Nepheline*, Sommite.—Occurs in small six-sided prisms, in lava ; colour white ; translucent ; splendent ; four cleavages ; fracture conchoidal ; when immersed in nitric acid, it becomes clouded and afterwards gelatinizes.—Before the blowpipe, on charcoal, its edges become rounded, and though it cannot be fused into a globule, it gives a blebby colourless glass. Melts with borax, quietly and slowly, into a transparent colourless glass.

638. Sp. gr. 2·3.—*Analcime*, Cubicite, Cubic Zeolite.—Occurs in the cavities of basaltic rocks, in small cubic or garnet-shaped crystals, generally aggregated ; pale grey ; transparent or translucent ; shining pearly ; becomes feebly electric by rubbing.—In the matrass, yields water and turns white. On charcoal, in a strong heat, becomes transparent, and melts without intumescence, into a diaphanous glass, slightly blebby. Fuses with borax, with much difficulty, even in powder ; forms a transparent glass, and leaves an opaque concrete residuum. Gives a transparent glass with soda, and a blue glass with solution of cobalt.

639. Sp. gr. 2·2.—*Stilbite*, Radiated Zeolite.—Occurs in masses composed of prismatic crystals, grouped in bundles, or diverging like a fan ; or in masses of laminæ closely adhering ; colours white, grey, and brown ; transparent ; translucent ; remarkably splendent pearly lustre ; hardness variable ; swells in acids, but does not gelatinize.—Behaves before the blowpipe like Prehnite, see 604, but emits a phosphoric light during its intumescence.

b. Heulandite, Foliated Zeolite. This mineral
has generally been considered a variety of Stilbite.
Occurs in prisms and in globular concretions; lustre
pearly and very splendent; yellowish white and
brownish red; translucent.—Before the blowpipe,
it melts with intumescence, during which it emits a
phosphoric light.

640. Sp. gr. 2·2.—*Laumonite,* Efflorescent Zeolite.
—Occurs in aggregated crystalline masses, deeply
radiated; greyish white; glistening pearly; hardness variable; effloresces, and becomes opaque and
tender on exposure to air; dissolves rapidly and
gelatinizes in muriatic acid.—Behaves before the
blowpipe like Prehnite, see 604, but first forms a
white globule of enamel, which ultimately becomes
a blebby translucent glass.·

641. Sp. gr. 2·2.—*Mesotype,* Needle Zeolite,
Needle-stone.—There are several varieties of this
mineral, the pyrognostic characters of which are as
follows:—In the matrass, alone, yields water. On
charcoal, radiated mesotype expands longitudinally,
and twists itself up like a screw, while the compact
variety intumesces, after which both melt into a
blebby colourless glass. The variety in large crystals
neither intumesces nor expands; but first becomes
opaque and then vitrifies. Melts with great difficulty with borax, into a transparent colourless glass.
Decomposes readily with mic. salt, producing a
glass which becomes opaline on cooling, and leaving
a skeleton of silica.

a. Crystallized Mesotype. Primitive form, a rectangular prism; secondary form, a long four-sided
prism, terminated by low four-sided pyramids;
structure lamellar; colour white, sometimes inclining to yellow, green, or red; transparent; shining
pearly. Electric by heat. Gelatinizes in acids.

b. Fibrous Mesotype, Radiated Needlestone. Occurs in globular concretions, composed of diverging
or stellular crystals or fibres, which are slender and

acicular, or flat and broad; sometimes found in distinct acicular crystals, radiating from a centre, and
so delicate as to resemble fine cotton. These occur
in the cavities of amygdaloid.

The mineral called *Thomsonite*, is a colourless
variety of radiated Needlestone.

Scolezite seems to be nearly allied to Thomsonite.

c. Mealy Zeolite, Pulverulent Mesotype. Occurs
in soft, dull, friable masses; having an earthy fracture, and a rough and meagre feel; greyish or reddish white; not pyro-electric.

d. Mesolite. Occurs massive, and in long and slender prisms, terminated by quadrilateral pyramids;
greyish or colourless; transparent; lustre shining
pearly.

FAMILY 2. *Softer than Fluor-spar.*

642. Sp. gr. 2·5.—2·0.—*Common Asbestus.*—Massive; structure parallel and curved fibrous or bladed; the fibres are coarser than those of Amianthus,
and scarcely flexible; colour dull green; lustre pearly; glistening; soft and unctuous. Melts before
the blowpipe, more easily than Amianthus, and forms
a greyish black globule of enamel.

643. Sp. gr. 2·3.—2·0.—*Amianthus,* Flexible Asbestus.—Occurs in very long and extremely slender
fibres, which are arranged parallel to each other
and are easily separated; remarkably flexible and
elastic; has the lustre and soft feel of silk; prevailing colour greenish white; slightly translucent.—
Before the blowpipe it phosphoresces; and in mass,
is fusible, with great difficulty, into a white enamel;
but, in filaments, it is easily melted, even by the
flame of a candle.

644. Sp. gr. 2·1.—*Ligniform Asbestus,* Mountain
Wood, Wood Asbestus, Rock-wood.—Occurs in
laminæ or plates, which have a promiscuous fibrous
structure like wood; tough; sectile; meagre; the

fibres are flexible and elastic; colour wood brown.
—Fusible into a black slag.

645. Sp. gr. 2·3.—*Gypsum*, Sulphate of Lime.—
In the matrass, alone, gives off water and becomes
milky white; after which, behaves like Anhydrite,
see 621.—There are the following varieties of this
mineral:—

   a. Crystallized Gypsum, Selenite. Occurs crystallized in flat oblique parallelopipedons; cleavage easily produces laminæ, which are thin and flexible,
but not elastic; colours white, yellow, and brown;
frequently colourless and highly transparent; lustre
shining pearly; very soft; yields to the nail.

   b. Fibrous Gypsum. Occurs massive, composed
of extremely delicate and nearly separate fibres,
either straight or curved; colour white and various;
lustre glistening pearly; translucent; cross fracture
lamellar, and very brilliant; soft. A very beautiful
mineral.

   c. Granular Gypsum. Massive, composed of an
aggregation of small crystalline laminæ, of which
the structure is lamellar, straight or curved; sometimes fibrous in its texture; shining pearly; translucent; very soft; generally white.

   d. Massive Gypsum, Alabaster. Massive; fracture
compact, passing to splintery; glimmering; very
soft; translucent on the edges; colour white, dotted
or veined with yellow or red.

   e. Earthy Gypsum. Earthy mass; white; dull;
friable.

646. Sp. gr. 2·2.—*Black Chalk*, Drawing Slate.
—Massive; slaty; black; dull; opaque; meagre;
stains paper black.—Acquires a superficial glazing
and a red colour before the blowpipe.

647. Sp. gr. 2·1.—*Clay.*—Plastic with water; more
or less unctuous to the touch; acquires a polish
from the nail.—Before the blowpipe; melts into a
slag.

   a. Earthy Clay, Common Brick Clay, Alluvial

Clay. Very plastic when pure; less so in proportion to the sand which is mixed with it.

_b._ Slaty Clay, Pipe Clay, Potters' Clay. Slaty; yields to the nail, but is scarcely plastic until it disintegrates by exposure to air.

GENUS 4. SPECIFIC GRAVITY LESS THAN 2·0., SOME SPECIES SUPERNATANT.

648. This Genus is not divided into Families. The minerals of which it is constituted are of various degrees of hardness;—some species being harder than fluor-spar; others extremely soft;—but they are very limited in number.

649. Sp. gr. 2·0.—1·2.—_Bole._—Amorphous; fracture conchoidal and glimmering; red and semi-transparent; grey and translucent on the edges; brownish black and opaque (_Mountain soap_); yields to the nail; streak shining; adheres to the tongue; breaks down in water.—Melts into a slag.

650. Sp. gr. 1·8.—_Pumice._—Massive; structure irregularly fibrous, with elongated cells; colour smoke grey; lustre shining pearly; fracture uneven glistening; translucent on the edges; harsh to the touch; harder than fluor-spar, but yields to the knife; sometimes swims on water.—Fusible before the blowpipe, into a dirty green blebby glass.

651. Sp. gr. 1·7.—_Fuller's Earth._—Massive; greenish brown; opaque; very soft; dull; fracture earthy uneven; unctuous; receives a polish from the nail; in water, becomes semi-transparent, falls to pieces, and forms a smooth pulp.—In the matrass, yields water and an empyreumatic odour, becomes clear, and then turns brown. On charcoal, heated gradually, it crackles; heated suddenly, it splits with violence; by a continued heat, it is melted into a white blebby glass. Fuses, with borax, into a transparent

colourless glass; with soda, into a globule of a grass green colour.

b. Lemnian Earth, Sphragid. This is probably a variety of Fuller's earth. It is yellowish grey, with ochreous spots. When immersed in water, it falls to pieces, evolving numerous air-bubbles. It is curious, that no mineralogist has stated the specific gravity of this substance.

652. Sp. gr. 1·0.—*Rock-Cork,* a variety of Asbestus.—Massive, composed of fibres interlaced; grey; opaque; dull; meagre; elastic; tough; yields to the nail; often supernatant.—Before the blowpipe, melts with difficulty into a white glass.

b. Mountain Leather. This is a variety of rock-cork, which occurs in thin flexible plates, having much the appearance of leather.

ORDER. 3.   Infusible before the Blowpipe, Alone.

653. This Order is divided into four Genera, as follows :—

GENUS 1. *Specific Gravity less than 5·0., but more than 3·0.*

GENUS 2. *Specific Gravity less than 3·0., but more than 2·5.*

GENUS 3. *Specific Gravity less than 2·5., but more than 2·0.*

GENUS 4. *Specific Gravity less than 2·0.— Some species supernatant.*

These Genera are all divided into Families, according to the different degrees of hardness of the minerals which they comprehend.

GENUS 1. SPECIFIC GRAVITY LESS THAN 5·0.,
BUT MORE THAN 3·0.

654. This Genus is divided into four Families :—

FAMILY 1. *Harder than Quartz—Scratch Quartz
with ease.*

FAMILY 2. *Rather harder than Quartz—Scratch
Quartz with some difficulty—Scratch Felspar
with ease.*

FAMILY 3. *Hardness equal to, or exceeding, that
of Felspar—Scratched by Quartz—Scratch Window-glass with ease.*

FAMILY 4. *Softer than Felspar—Scratch Fluor-
spar—Scratch Window-glass feebly.*

—

FAMILY 1. *Harder than Quartz—Scratch
Quartz with ease.*

655. Sp. gr. 4·7.—*Zircon.*—In grains and small
crystals, primitive form an obtuse octohedron, with
joints in two directions; colours various and pale;
translucent or transparent; refracts doubly. There
are three varieties; the pyrognostic characters of
which are as follow:—Before the blowpipe, per se,
the transparent colourless Zircon suffers no change;
but the red hyacinth loses its colour, and either becomes perfectly limpid, or assumes a slight tinge of
yellow; the brown opaque zirconite turns white and
gets full of cracks; and the blackish variety gives
off some water, becomes white, and apparently effloresces. But of all the varieties, none are fusible;
not even in powder or thin laminæ. Zircon fuses
readily into a clear glass with borax, which can be
made opaque by flaming; with mic. salt, it does
not give the least sign of fusion; neither does it melt
with soda. On platinum foil, it generally affords
a trace of manganese. The varieties follow :—

a. Hyacinth. Orange red; lamellar structure; cross

x

fracture conchoidal, with a vitreous lustre; semi-transparent.

b. Jargoon. Small prismatic crystals; transparent; yellowish, greyish or reddish smoky colour; also in round brown opaque masses.

c. Zirconite. Prismatic crystals; reddish brown; nearly opaque.

656. Sp. gr. 4·0.—*Corundum,* Common Corundum, Adamantine Spar.—Occurs massive, in rolled pieces, and in hexahedral prisms often bevelled; translucent; colour often greyish green, but various and dull; almost as hard as diamond.—Before the blowpipe, suffers no change, per se, whether tried in powder or fragment; but with borax, melts perfectly, though slowly, into a diaphanous glass, not capable of being made opaque by flaming. The powder fuses slowly with mic. salt, into a transparent glass. It does not melt with soda. On account of the alumina of which it is composed, it forms with solution of cobalt a beautiful dark blue colour; for this last experiment, the assay must be well powdered and the blast well kept up. These pyrognostic characters apply as well to the Sapphire and Ruby, as to Corundum.

657. Sp. gr. 4·0.—*Perfect Corundum,* Telesia.—Occurs in small rolled pieces, and in crystals; forms, a six-sided prism, or acute six-sided single or double pyramids; colours various; more or less transparent; possesses double refraction.—For its pyrognostic characters, see Corundum (656). There are two varieties:—

a. Sapphire. The hardest substance in nature next to the diamond; occurs in crystals that readily cleave in one direction; cross fracture conchoidal; blue, yellow, green, or colourless.

b. Oriental Ruby. Less hard than *a*, and more readily cleaved; structure lamellar; sometimes chatoyant; red, violet, or blue.

658. Sp. gr. 4·0.—*Emery,* a variety of Corundum.—Amorphous; granular; colour blackish and blue-

ish grey; aspect like that of a fine grained rock; lustre glistening; translucent on the edges.—Behaves like Corundum before the blowpipe.

659. Sp. gr. 3·8.—*Chrysoberyl,* Cymophane.— Occurs in rounded pieces and small crystals, primitive form a rectangular prism; cross fracture perfectly conchoidal, with a resino-vitreous lustre; colour light green, mixed with brown or yellow; sometimes shows an opalescent blueish light internally; semi-transparent; electric by friction.—Suffers no change, alone, before the blowpipe. Melts into a clear glass with borax, and remains transparent at every point of saturation. Complete fusion ensues with mic. salt, and a glass is formed which continues clear on cooling. Does not act with soda. An elegant blue colour is developed by a cake formed of solution of cobalt and powdered chrysoberyl, but no fusion takes place.

660. Sp. gr. 3·8.—*Pleonaste,* Ceylanite.—Colour black, but by transmitted light green or blue; translucent; in small crystals; form, the octohedron and its varieties; fracture flat conchoidal; lustre splendent.—Infusible, per se, even in powder. Forms a dark green iron-coloured transparent glass with borax. Intumesces with soda, and yields a black infusible scoria.

661. Sp. gr. 3·7.—*Spinelle Ruby,* Spinel.—In grains and crystallized; form, an octohedron, perfect or with edges replaced; red, violet, or yellow; fracture flat conchoidal; lustre splendent vitreous; softer than the oriental ruby.—Infusible, per se; but blackens and becomes opaque, then, on cooling, shows the following colours by transmitted light, fine chrome green, colourless, ruby tint. Melts slowly with borax; the result a glass transparent and nearly colourless. Fuses entirely with mic. salt if powdered. Intumesces with soda, but does not melt. Gives signs of manganese on platinum foil.

662. Sp. gr. 3·6.—*Topax.*—Occurs in long trans-

parent many-sided prisms, frequently with complicated summits; also, massive; colour various, prevailing yellow; fracture small conchoidal; lustre
splendent vitreous; softer than spinel; refracts
doubly; electric by heat with polarity.—Infusible,
alone, on charcoal, before the blowpipe; but in a
low heat some of the varieties change colour. At a
very high temperature, the striated sides of the
prisms become frosted with minute bubbles of gas.
Becomes white and opaque with borax, then melts
into a clear glass. Fuses slowly with mic. salt, and
leaves a skeleton of silica. Produces with cobalt
solution, a disagreeable blue colour.

663. Sp. gr. 3·5.—*Diamond.*—This mineral belongs to the combustible class, see 413, and is
merely repeated here because it agrees in many properties with the minerals of this Family. Hardness
superior to that of every other substance.—Infusible; but very slowly combustible at a white heat.

664. Sp. gr. 3·2.—*Andalusite.*—Massive, and in
slightly rhombic prisms; structure lamellar, with
rectangular joints; colour reddish; translucent.—
Infusible, alone, either in splinters or powder; but,
becomes covered with white spots. Fuses with borax,
into a transparent colourless glass.

FAMILY 2. *Rather Harder than Quartz—*
Scratch Quartz with some difficulty—Scratch
Felspar with ease.

665. Sp. gr. 3·6.—*Staurolite,* Staurotide, Grenatite-
—Form peculiar,—two six-sided prisms intersect each
other, either at right angles or obliquely; rarely in
single crystals; colour dark reddish brown; opaque
to translucent; fracture uneven; lustre glistening.—
Fragments are infusible alone but become of a dark
colour. Powder fuses on the edges into a black
scoria. Melts slowly with borax into a transparent
dark green glass. Does not melt with soda, but
effervesces and forms a yellow scoria. With mic.

salt fuses very slowly, leaving a little silica, and
forming a glass which is yellowish green and diaphanous while hot, but opaline and colourless when
cold.

666. Sp. gr. 3·5.—*Pycnite*, Schorlaceous Beryl.
—Occurs in long six-sided prisms, deeply striated
longitudinally; also, in parallel prismatic concretions, with transverse rents; yellowish and reddish
white; translucent; glistening; fragile; becomes
electric by heat.—Infusible before the blowpipe,
alone; but, in a strong heat, the longitudinal faces
of the crystal become covered with numerous small
white bubbles, in which respect it resembles the
topaz.

667. Sp. gr. 3·4.—*Pyrophysalite*, a variety of Topaz.—Occurs in large crystals, resembling those of
the topaz; also, in roundish masses; colour dull
greenish white; structure lamellar, and splendent
in one direction; fracture uneven and glimmering; translucent on the edges.—Infusible before
the blowpipe, but acquires a slight glazing, and disengages bubbles of gas.

668. Sp. gr. 3·1.—*Tourmaline*, Lyncurium of the An-
cients.—Occurs in rolled pieces and in prismatic crystals with many sides, deeply longitudinally streaked,
and having dissimilar terminations; principal colours
green and brown, but it is also white, yellow, red
and blue, never black, generally muddy; lustre splendent vitreous; semi-transparent; fracture conchoidal;
remarkably pyro-electric by heat, with polarity.—Before the blowpipe, *Red and Clear Green Tourmaline*,
alone, becomes white, intumesces, spits, assumes a
scoriaceous appearance, but does not melt. With
borax, after a slight effervescence, becomes white,
and fuses slowly into a colourless diaphanous glass.
Acts with mic. salt as with borax, but the glass becomes opaline on cooling. Melts difficultly with
soda into an opaque glass. Becomes dark green on
platinum foil. *Dark Blue Tourmaline* in large crys-

tals, (Indicolite,) swells to three times its original
size; the assay becomes curved and is converted into
a black scoria. Acts with the fluxes like the preceding varieties.

669. Sp. gr. 3·1.—*Rubellite*, Red Schorl, Red Variety of Tourmaline, Siberite.—Occurs in six, nine, or
twelve-sided prisms deeply longitudinally streaked;
colour red, with often a tinge of pink or violet.
—Alone, before the blowpipe, on charcoal, becomes
milk white, intumesces violently, splits obliquely,
does not fuse, but vitrifies on the edges. Effervesces
with borax and readily melts into a clear glass wherein
some flocculi may be observed to float and gradually
dissolve. Readily decomposes with mic. salt, after
effervescing, and forms an opaline glass, leaving a
silica skeleton. Fuses slowly with soda into an
opaque glass. Is much more soluble with the fluxes
than tourmaline. Exhibits intensely, on platinum
foil, the effects of manganese.

FAMILY 3. *Hardness about equal to, or exceeding, that of Felspar—Scratched by Quartz
—Scratch Window-glass with ease.*

670. Sp. gr. 4·7—4·3.—*Automalite*, Spinelle Zin-
cifere, Gahnite.—Dark blueish green octohedral crystals, nearly opaque.—Infusible alone. With borax
it fuses in very small quantity, even in powder,
forming a glass which is green while hot, but
colourless when cold.

671. Sp. gr. 4·2.—*Gadolinite.*—Occurs massive,
rarely crystallized; greenish black; fracture flat conchoidal; lustre splendent resinous; slightly translucent; affects the magnetic needle; forms a stiff grey
jelly when pulverized and digested in an acid. Varieties of this mineral behave differently before the
blowpipe:—*Expts.—a,* Gave off water in the matrass.
On charcoal, in a good flame, turned white and fused
quietly into a dark grey glass. With borax easily

formed a clear glass. With soda fused slowly into a red grey scoria. With mic. salt fused into a clear glass, leaving a skeleton of silica. The glass became opaline on cooling. On platinum foil gave traces of manganese.—*b*, In the matrass gave off no water nor volatile matter, but shone as if on fire, enlarged, and split. Did not fuse on charcoal, but exhibited the same phenomena as in the matrass; the particles which flew off during decrepitation glowed like sparks.—*c*, Intumesced alone, turned white, gave off moisture, and threw out cauliflower ramifications.—*b* and *c* behaved alike with the fluxes. With borax easily formed a glass which the reducing flame rendered bottle green. With mic. salt, only partially fused and formed no silica skeleton. On platinum foil, no indications of manganese were exhibited.

672. Sp. gr. 3·4.—*Chrysolite*, Peridot.—Occurs in angular and rolled pieces, and in prismatic crystals; fracture conchoidal; splendent vitreous; yellow mixed with green or brown; transparent; doubly refractive. —Infusible alone, but loses its transparency and becomes blackish grey. With borax it melts, without effervescence, into a transparent glass of a light green colour. Infusible both with mic. salt and soda.

673. Sp. gr. 3·4.—*Hyperstene*, see 574.

674. Sp. gr. 3·4.—*Allanite*.—Occurs massive, and in prismatic crystals; colour brownish black, when pulverized greenish grey; fracture small conchoidal; lustre shining, resino-metallic; opaque; brittle.—In the matrass gives off water, decrepitates, and becomes lighter in the colour. On charcoal, becomes greenish yellow on the surface, but does not fuse, even at the thinnest edges, in an intense and long-continued heat.

675. Sp. gr. 3·3.—*Anthophylite*.—Occurs crystallized and amorphous; structure radiated; colour reddish brown; lustre approaching to semi-metallic. —Infusible and unalterable both in fragment and

powder. Melts difficultly with borax into a glass tinged by iron. Slowly decomposes with mic. salt, leaving a skeleton of silica. Forms a scoriaceous mass with soda.

676. Sp. gr. 3·3.—*Zoisite.*—See 575.

677. Sp. gr. 3·2.—*Bronzite,* Diallage Metalloide.— Hardness variable. Yellowish or pinchbeck brown; lustre approaching to semi-metallic; occurs in distinct granular concretions; opaque in mass, but transparent in leaves; streak white.—Yields water, crackles, and assumes a clearer colour in the matrass; melts slowly on the edges into a grey scoria on charcoal. Melts with borax with much difficulty into a diaphanous glass tinged by iron.

678. Sp. gr. 3·1.—*Schiller Spar,* a variety of Diallage.—Hardness not uniform; colour various shades of green; splendent metallic; occurs in plates of different form; translucent in leaves; softer than bronzite.—Infusible before the blowpipe.

679. Sp. gr. 3·0.—*Jade.*—By some mineralogists erroneously said to be infusible. See 581.

680. Sp. gr. 3·0.—*Gehlenite.* Occurs in rectangular crystals; greenish grey; rough; dull; fracture uneven.—Alone, before the blowpipe, suffers no change; with borax melts with great difficulty into a glass faintly coloured by iron.

FAMILY 4. *Softer than Felspar—Scratch Fluor-Spar—Scratch Window-glass feebly.*

681. Sp. gr. 3·7—3·5.—*Sappare,* Disthene, Cyanite, Kyanite.—Hardness variable; colour different shades of blue; occurs amorphous with curved lamellar structure, or crystallized in long flat prisms; translucent; shining pearly; sometimes the colours are various and disposed in spots or stripes; some crystals, by friction, acquire negative electricity, others positive.—Infusible alone, in the strongest heat of the blowpipe, even in powder; but turns white. A blueish white splendent variety, called

Rhœtizite, changes to a red colour in a low heat, but in a higher, becomes white. Melts completely, though slowly, with borax, into a clear glass without colour. With a minute portion of soda, a partial fusion takes place, forming a blebby translucent globule, which, if gently heated by the outer flame, assumes a pale rose colour. With solution of cobalt, in a good heat, a beautiful deep blue colour is produced.

682. Sp. gr. 3·3.—*Coccolite*, Granular Augite.—In slightly coherent pea-like granular concretions; structure lamellar; various shades of green; shining.—Infusible before the blowpipe.

683. Sp. gr. 3·2.—*Olivine*, Granular Peridot, variety of Chrysolite.—In olive coloured semi-transparent masses, with often an iridèscent pseudo-metallic tarnish; found imbedded in basalt.—Before the blowpipe, turns brown on the edges but preserves its general colour and transparency, and does not melt. With borax forms slowly a clear glass, which does not become opaque by flaming. With soda is converted, with much labour, into a brown scoria.

———

GENUS 2. SPECIFIC GRAVITY LESS THAN 3·0., BUT MORE THAN 2·5.

684. This Genus is divided into three Families:—

FAMILY 1. *Hardness rather exceeding that of Quartz, or nearly the same—Not scratched by Quartz—Scratch Felspar.*

FAMILY 2. *Hardness about equal to that of Felspar—Scratched by Quartz—Scratch Window-glass with ease.*

FAMILY 3. *Scratched by Fluor-spar.*

———

FAMILY 1. *Hardness rather exceeding that of*

Quartz, or nearly the same—Not scratched by Quartz—Scratch Felspar.

685. Sp. gr. 2·7.—*Emerald.*—Vivid green.—Partially fusible. See 599.

686. Sp. gr. 2·7.—*Adularia.*—Partially fusible. See 602.

687. Sp. gr. 2·7.—*Prase*, a variety of Quartz.— Leek or olive green; considered to be quartz enclosing actynolite, distinct fibres of which are often seen disseminated within it. It is massive, translucent, and glistening.—Infusible per se; acts with the fluxes like quartz.

688. Sp. gr. 2·7.—*Cat's Eye*, a variety of Quartz.— In rounded pieces; grey and yellowish; translucent; lustre vitreo-resinous shining; exhibits a peculiar play of light, termed chatoyant, arising from the position of the fibrous amianthus enclosed in it; fracture small conchoidal.—It is infusible; but becomes opaque and spotted upon exposure to the blowpipe.

689. Sp. gr. 2·6.—*Beryl.*—Pale green.—Partially fusible. See 600.

690. Sp. gr. 2·6.—*Iolite.*—Partially fusible (601).

691. Sp. gr. 2·6.—*Chalcedony*, Flinty Quartz.— Never occurs crystallized; but botryoidal, stalactitic, nodular, in cubic pseudo-crystals, and as the petrifying matter of various organic remains; fracture waxy, compact, and flat conchoidal; has by transmitted light a cloudy or nebulous appearance, which is shaded by spots and stripes of various colours, chiefly blue, grey, or brown; lustre dull; chatoyant when polished in a certain direction.—Infusible before the blowpipe, but becomes opaque.

a. Carnelian.—Prevailing colour blood red, but sometimes white, yellow, or brown; in rough rounded pieces; lustre glistening; fracture perfect conchoidal. Sp. gr. sometimes 2·3.

b. Sard, a variety of chalcedony or carnelian.—

Of a deep rich orange yellow or brown colour; by transmitted light nearly blood red.

c. Onyx, is composed of alternate layers of brown and opaque white chalcedony.

d. Sardonyx, is composed of alternate layers of sard and onyx or milk-white chalcedony. It is a very beautiful mineral when cut and polished.

e. Mocha Stone, is a variety of chalcedony, containing arborizations, or vegetable filaments of various colours.

f. Heliotrope, Bloodstone. Dark green, with yellow and blood-red spots or stripes; translucent; lustre nearly resinous glistening. Sp. gr. sometimes 2·7.

g. Agate, is not a simple mineral, but a compound of various siliceous substances; its basis seems to be chalcedony; it occurs in pebbles composed of alternate concentric lamellæ of quartzose minerals. When cut and polished it assumes certain appearances which have acquired for it the names of Ribbon Agate, Brecciated Agate, Fortification Agate, Moss Agate, Petrifaction Agate, &c. If a thin slice of agate be held to the light it shows the substances of which it is composed very clearly. The red stripes if transparent are carnelian, if opaque jasper; the white stripes are opal; the blue stripes chalcedony; the broad transparent colourless stripes are quartz; the purple stripes amethyst. Agate pebbles of the most beautiful description abound in Scotland.

692. Sp. gr. 2·6.—*Flint*, Pierre à Fusil.—Occurs in nodules, forming the substance of certain marine organic remains, and in other particular shapes; colour grey passing into brown; more or less translucent; fracture perfectly conchoidal; internal lustre glimmering resinous; fragments sharp-edged; very easily frangible.—Infusible before the blowpipe, but turns white and opaque.

693. Sp. gr. 2·6.—*Lydian Stone*, Touchstone, variety of Flinty Slate.—Massive; black; opaque; traversed by quartz veins; when polished, used to

try the purity of gold and silver by the colour of a
streak left on it by those metals when drawn over
its surface.—Infusible.

694. Sp. gr. 2·6.—*Quartz.*—One of the most abundant of minerals. No other substance is found in
such a variety of colours, forms, and situations. It
occurs massive, in rolled pieces, and crystallized.
The primitive form of its crystal is the rhomboid;
but it commonly occurs in dodecahedrons formed of
two six-sided pyramids joined base to base; sometimes the pyramids are separated by a six-sided
prism of which the alternate angles are frequently
replaced. Fracture conchoidal or splintery; lustre
vitreous splendent to glimmering; highly transparent to nearly opaque; causes double refraction when
very transparent; scratches glass; does not yield to
the knife.—Two pieces rubbed against each other
are phosphorescent, and exhale an odour resembling
that of the electric fluid. Infusible before the blowpipe. Insoluble in all acids but the fluoric. Several varieties of this mineral we have described in
other places; the three following only remain to be
described here.

a. Rock Crystal.—This is the purest and most
transparent variety of quartz: occurs in rolled pieces
and crystallized in the forms above described; the
prisms being transversely striated; lustre splendent;
finely doubly refractive; frequently without the
least colour, but sometimes coloured by some accidental ingredient. The stone called *cairngorum* is
a smoke coloured variety.

b. Common or Amorphous Quartz. — Occurs
massive, in grains, in rolled pieces, (pebbles or
chucky stanes,) in other particular shapes, and
in crystals; colour various shades of white, grey,
brown, yellow, green, and red; lustre of the crystals shining, of the rolled pieces glimmering; fracture splintery or parallel fibrous; translucent; fragments sharp-edged. *Hyacinths of Compostella* are

orange-coloured quartz crystals. *Rose Quartz* is simply massive quartz, semi-transparent, and of a pale rose-red colour. *Milk Quartz* is a blueish-white variety of massive quarts. *Fat Quartz* is so named because the fractured surface appears as if it had been rubbed with oil. *Ferruginous Quartz* or Iron Flint, is quartz coloured red by about 5 per cent. of iron.

c. Amethyst, Violet Quartz.—This is quartz having a violet-blue or amethystine colour. It occurs amorphous, but commonly in imperfectly formed pyramidal crystals often radiated; found in veins and in the hollow cavities of agates; transparent; the colour often not uniform throughout the crystal. The massive variety composed of densely-aggregated imperfect prisms, which give to the concretions the appearance of a coarsely fibrous structure.

695. Sp. gr. 2·6.—*Common Hornstone*, Chert.— Amorphous, in nodules, never crystallized; colour usually grey; fracture splintery or conchoidal; dull; slightly translucent.—Looks like compact felspar, but that mineral is fusible, while this is infusible.

b. Woodstone, is wood converted into hornstone by petrifaction, often preserving its fibrous appearance.

696. Sp. gr. 2·6.—*Hornstone-Slate*, Flinty-Slate, Indurated Slate.—Massive; colour grey; sometimes striped; structure slaty; opaque or translucent on the edges; dull; scarcely so hard as quartz.—Infusible per se.

697. Sp. gr. 2·6.—*Egyptian Jasper.*—In loose rounded masses with a rough surface; colours red, brown, and yellow, singularly arranged in curved and contorted stripes; structure compact; fracture conchoidal; takes a high polish when cut; rather harder than quartz.—Infusible per se.

a. Brown Egyptian Jasper.—Yellow, with brown concentric delineations and black spots; translucent on the edges.

Y

b. Red Egyptian Jasper.—Between scarlet and blood red, with ring-shaped delineations ; opaque.

698. Sp. gr. 2·5.—*Plasma*, Green Flint.—Occurs in angular fragments among the ruins of Rome ; dark-green with white and yellow dots ; translucent. —Infusible.

FAMILY 2. *Hardness about equal to that of Felspar—Scratched by Quartz——Scratch Window-glass with ease.*

699. Sp. gr. 2·9.—*Chiastolite*, Macle, Hollow Spar.— Occurs in long slender quadrangular prisms of a greyish white colour, each enclosing within it a dark blue or black prism of a similar form with the exterior ; the white part is laminar ; glistening ; translucent.— Whitens in the blowpipe flame, but does not melt. A very thin cake made of the powdered mineral, concretes into a mass. Melts difficultly with borax into a clear glass. Gives a blue colour with solution of cobalt. The black part affords, alone, a black glass.

700. Sp. gr. 2·7.—*Chrysoprase.*—A beautiful variety of Chalcedony ; colour apple green of various degrees of intensity ; highly translucent ; its other characters agree with chalcedony, except that it is softer.—Before the blowpipe, it does not melt, but becomes white and opaque.

701. Sp. gr. 2·7.—*Indianite.*—A rare mineral ; amorphous ; greyish white ; granular ; laminar ; translucent ; less hard than felspar ; scratches window-glass ; softened by acids.—Infusible before the blowpipe.

702. Sp. gr. 2·5.—*Common Jasper.*—Occurs in veins ; amorphous ; colours yellow, brown, and red, of various shades, frequently intermingled ; opaque ; generally dull ; fracture conchoidal or even ; often brittle ; sometimes glistening internally.—Infusible before the blowpipe.

b. Striped Jasper, Ribbon Jasper.—Differs from common jasper in the arrangement of its colours, which are disposed in spots, stripes, clouds, or concentric curves. Some specimens are extremely beautiful.

FAMILY 3. *Scratched by Fluor-spar.*

703. Sp. gr. 2·9.—*Pot-stone*, a variety of Serpentine.—Amorphous; structure undulatingly slaty; greenish grey; glistening; translucent; yields to the nail; unctuous; difficultly frangible; sectile.—Infusible before the blowpipe. In some countries, this mineral is turned on a lathe into culinary vessels which resist the action of fire.

704. Sp. gr. 2·9.—*Pinite.*—Nearly infusible (622).

705. Sp. gr. 2·8.—2·5.—*Pearl-spar*, Brown Spar.—Soluble in acids. See 557.

706. Sp. gr. 2·8.—2·7.—*Granular Limestone*, Statuary Marble.—*Compact Coloured Marble.*—*Calcareous Spar.*—Effervesce with acids. See 550 *g*, *h*, *a.*

707. Sp. gr. 2·8.—*Agalmatolite*, White Talc.—Occurs massive; structure imperfectly slaty; greenish grey veined with brown or blue; lustre glimmering, greasy; translucent; unctuous; yields to the nail.—Alone, in the matrass, gives off water having an empyreumatic odour, and blackens. Before the blowpipe, on charcoal, turns white, and presents at the extremity of the projecting part, some marks of fusion. Gives a colourless glass with borax.

708. Sp. gr. 2·8.—*Talc.*—Occurs in hexagonal plates, and massive; structure finely and curvedly laminar; the laminæ are easily separable from one another; translucent; flexible, but not elastic; prevailing colours silver white and green; lustre splendent pearly; soft; very unctuous; sectile; leaves a pearly white streak when rubbed on paper.—Specimens from different localites, vary considerably in their pyrognostic habitudes. In general, Talc exfoliates and whitens before the blowpipe, per se, but

does not fuse. Some varieties, however, give with difficulty a very minute globule of enamel.

a. Crystallised Talc, Venetian Talc. Occurs in regular minute six-sided tables of a white or light green colour.

b. Massive Talc. This is less flexible and translucent than the preceding; colour apple green; structure often radiated.

c. Indurated Talc. Massive; greenish grey; structure fibrous and curvedly slaty; shining pearly lustre; somewhat translucent. Sp. gr. 2·9..

709. Sp. gr. 2·8.—*Asbestous Tremolite.*—Hardness variable. See 571 *b.*

710. Sp. gr. 2·7.—*Magnesite,* Compact Carbonate of Magnesia.—Amorphous, tuberose, and spongiform; fracture splintery or flat conchoidal; colour yellowish grey, with dots or dendritic delineations of blackish brown; dull; nearly opaque; yields to the nail; meagre; adheres to the tongue; dissolves in sulphuric acid, and affords crystals of sulphate of magnesia.—In the matrass, gives off hardly any water. Crackles, shrinks, and hardens on charcoal, and then acts on moistened brazil-wood paper, like an alkali.

711. Sp. gr. 2·7.—*Roestone.—Dolomite.—Chalk.—* Soluble in muriatic acid. See 550 *i*; 556, and 550 *l.*

712. Sp. gr. 2·7.—*Steatite.*—Massive, sometimes with pseudo-morphous crystals of the same substance imbedded; fracture splintery; unctuous; yields to the nail; but does not adhere to the tongue; colours pale grey, yellow, or red; dull; translucent on the edges; sectile, cutting with a smooth shining surface.—It hardens before the blowpipe, and turns black, but is not fusible.

713. Sp. gr. 2·6.—*Alum-stone.*—Massive; reddish white; translucent; lustre dull; fracture earthy uneven.—Alone, in the matrass, it decrepitates, and yields a sulphureous gas. On charcoal, in a good ', it contracts but does not fuse. With borax, it

effervesces and melts into a transparent glass free
from colour. Does not fuse with soda. Gives a fine
blue glass with solution of cobalt.

714. Sp. gr. 2·6.—*Slate Spar.*—Effervesces violently in acids. See 550 *b*.

. 715. Sp. gr. 2·5.—*Serpentine.*—This is usually described as infusible; but Berzelius found it partially
fusible. See 630.

Genus 3. Specific Gravity less than 2·5.,
but more than 2·0.

716. This Genus is divided into three Families:—

Family 1. *Hardness varying from that of Quartz*
to that of Felspar.

Family 2. *Scratched by Felspar—Scratch Window-*
glass and Fluor-spar.

Family 3. *Softer than Fluor-spar.*

Family 1. *Hardness varying from that of*
Quartz to that of Felspar.

717. Sp. gr. 2·4.—*Leucite,* Amphigène, White
Garnet.—Occurs in little rounded masses, also in
crystals whose planes are 24 equal and similar tra-
peziums; structure lamellar; colour greyish white;
translucent; shining vitreous; scratches glass with
difficulty.—Infusible before the blowpipe, alone,
even in powder; but it melts if mixed with pulverized carbonate of lime. Fuses slowly with borax
into a clear glass. Effervesces with soda, and forms
a blebby glass.

718. Sp. gr. 2·4.—*Sodalite.*—Massive, and in rhombic dodecahedrons; colour blueish green; cleavage
twofold; fracture conchoidal; lustre shining resi-.
nous; translucent.—Before the blowpipe, on char-:
coal, Berzelius found one specimen   se, with very

brisk intumescence, into a distorted colourless glass; another, in like circumstances, suffered no change, excepting that, by a powerful blast, the edges became rounded.

719. Sp. gr. 2·2.—*Wavellite*, Sub-Phosphate of Alumina, Hydrargillite.—Occurs in mammillated or hemispherical concretions, varying in size from that of a pea to a walnut; the concretions being composed of delicate acicular prisms, closely adhering, and elegantly radiating from the centre; prevailing colour greenish white; translucent; shining vitreous. If a fragment be laid on a watch glass with a drop of sulphuric acid, and heated, the glass becomes slightly corroded, from the disengagement of fluoric acid.—In the matrass, alone, water is given off, the drops of which, towards the end, are loaded with fluoric acid, as may be known by the usual signs. On charcoal, it intumesces, loses its crystalline form, becomes snow white and opaque, but does not melt. Treated with boracic acid and iron, it gives a fused regulus of phosphuret of iron.—The following substances are said to be varieties of Wavellite :—

b. Diaspore. Occurs in curvilinear lamellæ, easily separable; having a shining pearly lustre; and a sp. gr. of 3·4.—A fragment held in the flame of a candle, explodes, and separates into minute particles.

c. Turquoise. An opaque greenish-blue stone employed in jewellery.

720. Sp. gr. 2·2.—*Menilite*, a variety of Semi-Opal. —Occurs in tuberose pieces; grey and brown; translucent on the edges; structure slaty. Insoluble in all acids but the fluoric.—Infusible before the blowpipe.

721. Sp. gr. 2·2.—*Cacholong*, a variety of Chalcedony.—Occurs in loose masses; colour milk white; : brittle; fracture flat conchoidal; somewhat nt; externally dull; rare.—Infusible before pipe.

722. Sp. gr. 2·1.—*Jasper-Opal.*—Massive; nearly opaque; red and yellow; lustre vitreous shining.—Infusible before the blowpipe. Distinguished from the other varieties of jasper by its more easy frangibility, and its low specific gravity.

FAMILY 2. *Scratched by Felspar—Scratch Window-glass and Fluor-spar.*

723. Sp. gr. 2·4.—*Hyalite*, Muller's Glass.—Bears a striking resemblance to gum arabic; occurs lining the cavities of basaltic amygdaloid; pale yellow; shining vitreous; semi-transparent; smooth and mammillated; fragile.—Infusible before the blowpipe.

724. Sp. gr. 2·2.—*Semi-Opal.*—Nearly opaque; has no play of colours; lustre faint and resinous; colours white, grey, and various; harder than Opal.—Infusible before the blowpipe.

b. Wood Opal. Petrified wood penetrated by semi-opal; remarkable for its ligneous structure; distinguished from wood-stone (wood petrified by chert, &c.) by its superior lightness, translucency, and conchoidal fracture.

725. Sp. gr. 2·1.—*Opal.*—Varieties of this mineral, having properties peculiar to themselves, are described elsewhere: we have only to notice four in this place :—

a. Precious or Noble Opal. Occurs most frequently in nodules, never crystallized; colour blueish white by reflected light; pale orange by transmitted light; but chiefly remarkable for the beautiful chatoyant or opalescent appearance it possesses; highly translucent; internal lustre splendent vitreous; brittle; fracture conchoidal. Insoluble in all acids except the fluoric.—It is infusible before the blowpipe, but it decrepitates, loses its colour, and becomes opaque.

b. Common Opal is of various shades of white, green, yellow, and red; but is entirely without the play of colours which characterizes the noble opal,

and is less transparent than that variety. In its
other characters, however, it is much the same.

c. Milk-Opal, Opal-Cacholong. Translucent and
milk white.

d. Fire Opal. Differs only from the noble opal,
with which it sometimes occurs, in possessing merely
a red reflection when turned toward the sun.

FAMILY 3. *Softer than Fluor-spar.*

726. Sp. gr. 2·2.—*Porcelain Earth*, Kaolin, Porcelain Clay, Disintegrated Felspar.—Amorphous;
white of various tints; soft but not unctuous;
slightly cohesive.—Infusible before the blowpipe.

727. Sp. gr. 2·2.—*Cimolite*, a variety of Fuller's
Earth.—Massive; structure slaty; greyish white;
dull; opaque; fracture earthy uneven; yields to the
nail; adheres to the tongue; sectile; tough; separates in water into thin slaty laminæ, which, by
trituration, form a soft pulp.—Infusible.

728. Sp. gr. 2·2.—*Lithomarge.*—There are two
varieties of this mineral. Some specimens phosphoresce when heated; others, when moistened, give
out an agreeable smell like that of nuts.—Before the
blowpipe, it is infusible:—

a. Friable Lithomarge. White scaly particles
slightly cohering; soils; greasy; adheres to the
tongue; phosphoresces.

b. Indurated Lithomarge. Amorphous; colours
various; mottled; dull; opaque; fracture earthy;
streak shining; yields to the nail; greasy; heavier
than the preceding.

729. Sp. gr. 2·2.—*Tripoli.*—Massive; fracture
coarse dull earthy, structure slaty; yellowish grey;
dull; opaque; meagre; rough; yields to the nail;
does not adhere to the tongue.—Infusible.

730. Sp. gr. 2·1.—*Adhesive Slate.*—Amorphous;
grey; dull; opaque; fracture slaty in the large,
earthy in the small; absorbs water with avidity,

air-bubbles separating with a bubbling noise.—Infusible.

731. Sp. gr. 2·1.—*Native Magnesia,* Hydrate of Magnesia.—Occurs in small veins, in serpentine; rare; structure broadly fibrous, radiated; colour white; lustre shining pearly; translucent in mass, transparent in foliæ; soft; elastic; adheres to the tongue. Soluble in acids.—In the matrass, gives off water. Alone, on charcoal, swells, crackles, and turns milk white; but does not fuse.

732. Sp. gr. 2·0.—*Indurated Clay,* Fire Clay, Stourbridge Clay.—Amorphous; fracture earthy granular; grey or brown; hardness variable; by exposure to the air rendered soft and plastic.—On charcoal, in a low heat, turns white; in a strong heat, becomes scoriaceous but does not fuse.

GENUS 4. SPECIFIC GRAVITY LESS THAN 2·0.; SOME SPECIES SUPERNATANT.

733. This Genus is not divided into Families.

734. The minerals of which it is constituted are of various degrees of hardness,—some species being harder than Fluor-spar, others extremely soft,—but they are very limited in number.

735. Sp. gr. 1·9.—0·5.—*Polishing Slate.*—Amorphous; slaty; grey and yellow, in stripes; dull; opaque; very soft. Sp. gr. 0·5., but when it has imbibed water, 1·9.—Infusible.

736. Sp. gr. 1·8.—*Siliceous Sinter.*—This term denotes a kind of siliceous concretion. There are three varieties:—

a. Common Siliceous Sinter. Greyish, reddish, and brownish white; occurs stalactitic and in various particular forms, sometimes enclosing plants; dull or glistening; fracture conchoidal or fibrous;

generally porous; sometimes translucent on the edges; very brittle.

b. Opaline Siliceous Sinter. White, with dark spots and lines; massive, and in distinct concretions; glistening; adheres to the tongue; resembles opal.

c. Pearl Sinter, Fiorite. Occurs stalactitic and in botryoidal concretions composed of thin concentric laminæ; white or grey; pearly; lustre shining to dull; translucent; scratches glass and fluor-spar, but is scratched by quartz; brittle.—Sp. gr. 1·9.

737. Sp. gr. 1·7.—*Aluminite*, Pure Clay, Sub-Sulphate of Alumina.—Occurs in reniform pieces; yellowish white; dull; opaque; fracture earthy; yields to the nail; almost friable; meagre; strongly adheres to the tongue.—On charcoal, alone, infusible. Gives off water and sulphurous acid, in the matrass.

738. Sp. gr. 1·6.—*Meerschaum*, Sea Foam, Earthy Carbonate of Magnesia, Ecume de Mer.—Amorphous; colour yellowish white; opaque; dull; fracture fine earthy; yields easily to the nail; adheres strongly to the tongue; often very porous and supernatant; unctuous; lathers with water like soap.—In the matrass, yields water and an empyreumatic odour, and turns black. On charcoal, regains its white colour, contracts, and on the thinnest edges shows some symptoms of a white enamel. Assumes, with solution of cobalt, a fine lilac colour.

Class 4. SALINE MINERALS.

739. For the *Essential Characters* of this Class, see paragraph 392.

It is divided into two Orders, as follows:—

Order 1. *When Dissolved in Water, afford a Precipitate with Carbonated Alkali.*

ORDER 2. *When Dissolved in Water, do not afford a Precipitate with Carbonated Alkali.*

These Orders are not divided into Genera, because the minerals of which they are constituted are few in number, and easily discriminated.

ORDER 1. When Dissolved in Water, afford a Precipitate with Carbonated Alkali.

740. *Sulphate of Magnesia,* Epsom Salt.—Occurs in crystalline fibres; rarely pulverulent; greyish white; transparent; opaque; soft; brittle; taste peculiarly bitter and saline; seldom found in a solid state.—Before the blowpipe, dissolves very easily by means of its water of crystallization, but when dried, it is difficultly fusible.

741. *Native Alum.*—Occurs as an efflorescence on argillaceous minerals; also, stalactitic, in delicate capillary crystals, and massive, with a fibrous texture, and silky lustre; yellowish white; to the taste sweetish, styptic, and acidulous; harder than gypsum. Sp. gr. 1·8.—It melts easily before the blowpipe, by means of its water of crystallization, and by the continuance of heat, is converted into a white spongy mass.

742. *Green Vitriol,* Sulphate of Iron.—Occurs massive, stalactitic, disseminated, and crystallized in acute rhomboids. It arises from the decomposition of Iron Pyrites. Its colours are various shades of green and yellow. Rare. Solutions of this salt are turned black by solution of galls, and blue by solution of prussiate of potass. Sp. gr. 2·0.

743. *Blue Vitriol,* Sulphate of Copper.—Occurs massive, stalactitic, pulverulent; colour fine bright blue or blueish green; to the taste nauseous, bitter, metallic; a portion dissolved in a drop of water, and

spread on the surface of iron, immediately covers it
with a film of copper. Sp. gr. 2·2.

744. *White Vitriol*, Sulphate of Zinc.—Occurs
amorphous, and in various particular forms; colour
yellowish white; fracture fibrous radiated; shining;
translucent; soft; brittle; to the taste nauseous
metallic. Sp. gr. 2·0.—Before the blowpipe, it fuses
with ebullition, giving off a large quantity of sulphurous acid, and leaving a grey scoria.

745. *Red Vitriol*, Sulphate of Cobalt. — Occurs
stalactitic, massive, and investing; surface furrow-
ed; colour pale rose red; more or less transparent;
crystalline; styptic taste; its solution affords, with
carbonate of potass, a pale blue precipitate, which
tinges borax of a pure blue colour.

746. *Muriate of Mercury*, Horn Mercury.—A
metallic mineral, see 431. Dissolves in water, and
affords with lime water, a precipitate of an orange
colour.

ORDER 2. When Dissolved in Water, do not
afford a Precipitate with Carbonated Alkali.

747. *Native Boracic Acid*, Sedative Salt, Sassolin.
—Massive; friable; composed of minute white
pearly scales, which adhere somewhat to the fingers;
bitter and sub-acid to the taste; very light.—Very
easily fusible before the blowpipe, with slight intumescence, into a transparent globule. If placed
on wetted brazil-wood paper, it bleaches it.

748. *Borax*, Borate of Soda, Tincal.—Occurs in
prismatic hexagonal crystals, compressed and variously terminated; primative form, an oblique rhombic prism; colour white, with a tinge of blue or
green; shining resinous; semi-transparent; soft;
brittle; refracts doubly. Sp. gr. 1·6.—Before the
blowpipe, alone, it intumesces with violence, car-

bonizes, yields an empyreumatic odour, and melts into a transparent colourless globule.

749. *Natron*, Carbonate of Soda, Trona.—Massive, fibrous, in crusts, and efflorescent; yellowish white; to the taste urinous and saline; translucent or opaque; the fibrous variety consists of an' aggregation of minute shining vitreous crystals, often radiated. Effervesces violently with acids. Sp. gr. 1·4.—Very easily fusible before the blowpipe.

750. *Glaubersalt*, Sulphate of Soda.—Occurs forming efflorescent incrustations; yellowish or greyish white; to the taste, cooling, bitter, and saline.—In the matrass, gives off so much water of crystallization as to dissolve itself; the salt then dries, and on charcoal, melts, is absorbed, and converted into a sulphuret. With soda, it penetrates the charcoal, by which it is distinguished from salts which have earthy bases.

751. *Nitre*, Nitrate of Potass, Saltpetre.—Occurs in crusts, and in groups of capillary crystals; colour yellowish white; to the taste cooling and saline; translucent; brittle. Sp. gr. 2·0.—In the matrass, yields water, and melts at a very low heat; on charcoal, at the moment of fusion, it detonates, leaving an alkaline mass.

752. *Sal-Ammoniac*, Muriate of Ammonia.—Occurs massive, with a fibrous structure, plumose, in crusts, and in minute octohedral crystals; white when pure, but often coloured by accidental ingredients; transparent to opaque; externally dull, but shining vitreous internally; to the taste peculiarly pungent and acrid; when moistened, and rubbed with quicklime, it gives out a pungent ammoniacal odour.—Sp. gr. 1·6.

753. *Common Salt*, Rock Salt, Muriate of Soda.—Occurs in beds, in large columnar or spheroidal concretions, and crystallized in cubes and octohedrons; structure obscurely lamellar; breaks readily into cubic fragments, the cube being the primitive form;

when pure of a white colour, but when mixed with
accidental ingredients, of various colours; lustre
shining vitreous; transparent to opaque; readily
yields to the knife; deliquesces in the atmosphere;
taste same as that of common culinary salt. Sp. gr.
2·2.—In the matrass, decrepitates and yields water.
On charcoal, melts, disengages fumes, and is absorbed. With mic. salt and oxide of copper, produces the fine blue flame which characterises muriatic acid.

b. Fibrous Rock Salt. Colour greyish white,
sometimes striped with red or blue; massive and in
concretions, the structure of which is rather fine,
and generally waved fibrous; fragments splintery;
semi-transparent.—It decrepitates briskly before the
blowpipe, or when laid on burning coals.

754. *Glauberite.* — Crystallized in oblique flat
rhomboidal prisms; striated laterally; pale yellow
or grey; translucent; harder than gypsum; softer
than calcareous spar; structure lamellar. Sp. gr. 2·7.
—When immersed in water it becomes opaque, and
partially dissolves.—In the matrass, decrepitates violently, and yields a little water; afterwards, at a low
red heat, melts into a transparent glass. On charcoal, it turns white on the first impulse of the flame;
then melts into a clear globule, which changes, on
cooling, into a white enamel.

MISCELLANEA.

LAVOISIER'S EXPERIMENTS WITH THE OXYGEN GAS BLOWPIPE.

755. *Rock Crystal*, or pure siliceous earth, is infusible, but becomes capable of being softened or fused when mixed with other substances.

756. *Lime*, *Magnesia*, and *Barytes*, are infusible, either when alone, or when combined together; but, especially lime, they assist the fusion of every other body.

757. *Argill*, or pure base of alum, is completely fusible, per se, into a very hard opaque vitreous substance, which scratches glass like the precious stones.

758. All the compound *Earths* and *Stones* are readily fused into a brownish glass.

759. All the *Saline* substances, even fixed alkali, are volatilized in a few seconds.

760. *Gold*, *Silver*, and probably *Platinum*, are slowly volatilized without any particular phenomenon.

761. All other *Metallic Substances*, except mercury, become oxidated, though placed upon charcoal, and burn with different coloured flames, and at last dissipate altogether.

762. The *Metallic Oxides* likewise all burn with flames. This seems to form a distinctive character for these substances, and even leads me to believe, as was suspected by Bergman, that barytes is a metallic oxide, though we have not hitherto been able to obtain the metal in its pure or reguline state.

763. Some of the *Precious Stones*, as rubies, are capable of being softened and soldered together, without injuring their colour, or even diminishing their weights. The hyacinth, though almost equally

fixed with the ruby, loses its colour very readily. The Saxon and Brazilian topaz, and the Brazilian ruby, lose their colour very quickly, and lose about a fifth of their weight, leaving a white earth, resembling white quartz, or unglazed china. The emerald, chrysolite, and garnet, are almost instantly melted into an opaque and coloured glass.

764. The *Diamond* presents a property peculiar to itself; it burns in the same manner as combustible bodies, and is entirely dissipated.

The series of experiments which produced the above results was made in the years 1782-3. We beg to point, in an especial manner, the reader's attention to the sagacious conjecture contained in paragraph 762, the subsequent demonstration of the truth of which, by Sir Humphrey Davy, was one of the things that obtained for him so great a reputation.

PROFESSOR HARE'S EXPERIMENTS WITH THE OXY-HYDROGEN BLOWPIPE.

765. *Silica*, finely powdered and moistened with water.—Fusion perfect; result, a colourless glass.

766. *Alumina.*—Fusion perfect; result, a milk white enamel.

767. *Barytes.*—Fusion immediate; accompanied by intumescence, produced by the liberation of water; after which the assay became solid and dry, but soon melted again; result, a perfect globule of greyish white enamel.

768. *Strontia.*—Behaved like barytes.

769. *Glucina.*—Fusion perfect; result, a white enamel.

770. *Zirconia.*—Behaved like glucina.

771. *Lime.*—Produced a light, the splendour of which was insupportable to the naked eye. When the assay was viewed through deep-coloured glasses,

(as all experiments of this kind ought to be,) it was seen to become rounded at the angles, and sink gradually, till there only remained a small globular protuberance. The surface of this protuberance was converted into a perfectly white and glistening enamel, in which a lens discovered a few minute pores, but not the slightest earthy appearance.

772. *Magnesia.*—The escape of water caused the vertex of the cone of magnesia repeatedly to fly off in flakes; the top of the frustrum that thus remained, gave a reflection of light nearly as powerful as that given by lime.—After a few seconds, the assay, being examined by a magnifying glass, no rough nor earthy particles could be perceived, but a surface of perfectly white smooth enamel.

773. *Gun Flint.*—Fusion rapid; attended by ebullition and the separation of numerous small ignited globules, which appeared to burn away, as they rolled out of the current of flame; result, a splendid and beautiful enamel.

774. *Chalcedony.*—Fusion rapid; result, a beautiful blueish white enamel, resembling opal.

775. *Oriental Carnelian.*—Fusion attended with ebullition; result, a semi-transparent white globule, possessing a fine lustre.

776. *Red Jasper*, from the Grampians.—Fusion slow; attended by a slight effervescence; result, a greyish black slag, with white spots.

777. *Beryl.*—Fusion instantaneous; attended with violent ebullition during exposure to the flame, which was renewed upon its re-exposure after cooling; result, a vitreous globule of a beautiful blueish white colour.

778. *Emerald of Peru.*—Behaved like the beryl; but the resulting globule was green, and perfectly transparent.

779. *Leucite.*—Fusion instantaneous; attended with strong ebullition, during which numerous ignited globules darted from the assay, and either flew

z 2

in the air, or rolled upon the charcoal, and burnt; result, a perfectly transparent white glass.—Very probably, as leucite contains a large proportion of potass, these globules were potassium.

780. *Other Substances which were easily fused.*— Porcelain, common pottery, fragments of hessian crucibles, various kinds of clay, Wedgewood's-ware, fire-brick, and several compound rocks.

DR. CLARKE'S EXPERIMENTS WITH THE OXY-HYDROGEN BLOWPIPE.

781. The series of experiments made by Dr Clarke with the gas blowpipe, was the most important which has ever been made on mineral bodies exposed to so high a temperature. In a work like the present it is, therefore, but right to bestow a considerable degree of attention upon these experiments. We have consequently drawn up a short description of the most remarkable phenomena attending the fusion of various bodies which were tried, and of the results which that fusion produced. The length of this account, much as we have studied brevity in its composition, incroaches considerably on the limits of our little volume; yet we consider it of too important a nature to be farther shortened or omitted.—The student will acquire a fund of useful information by the comparison of these experiments with those described in the preceding pages as made with the mouth blowpipe; the results in the two cases, when a mineral is assayed with the gas and with the common blowpipe, being sometimes exceedingly curious. If he possesses a gas blowpipe, this account will serve as a manual whereby he may direct his operations; and should he be desirous of acquiring more extensive information on the subject, he has only to consult " *The Gas Blowpipe, or the Art of Fusion by Burning the Gaseous constituents of Water,*" published by Dr Clarke about eight years ago.

782. The following circumstances relative to the experiments, require to be borne in mind.—1. Many of the substances exposed to the gas blowpipe, being in the form of powder, were mixed into a paste with olive-oil, previous to the assay.—2. The results of the experiments were cut with a file, to discover the metallic lustre.

783. Pure Oxide of Calcium, Lime.—Fusion very difficult; result, a substance with a limpid botryoidal glassy upper surface; under surface black. It was supported in a platinum crucible.

784. The fusion of lime and all its compounds by the Gas Blowpipe, is accompanied by a beautiful lambent flame of an amethystine hue.

785. Crystallized Carbonate of Lime, Iceland Spar. — Fusion extremely difficult; result, a brilliant limpid glass.

786. Hydrous Carbonate of Lime, Arragonite.— Fusion difficult; result, same as that of 783.

787. Common Chalk, Carbonate of Lime.—Fusion easy; result, yellowish grey enamel; yielding to clear pearly glass.

788. Oolite, Pure Carbonate of Lime, Ketton Stone.—Fusion very difficult, attended with dense white fumes; result, a yellowish grey glass, mammillated.

789. Compact Transition Limestone, Limestone of Parnassus.—Fusion difficult; result, a white milky enamel.

790. Primary Foliated Limestone, variety of Parian Marble.—Fusion very difficult, attended with a deposition of white oxide; result, a snow white enamel, edges opalescent.

791. Limestone from the summit of the principal Pyramid of Egypt.—Fusion attended with dense white fumes; result, a white frothy enamel, full of bubbles.

792. Matrix of the Human Skeleton discovered at Guadaloupe, Calcareous Tophus, Tufaceous Limestone.—Fusion difficult, attended with a deposition of white oxide and intumescence; result, an intensely white enamel.

793. Marble from Tivoli, exceedingly Compact Limestone. — Fusion extremely difficult, attended with a deposition of white oxide; result, a snow white enamel.

794. Marble from Pompeii, Compact Granular Limestone.—Fusion less difficult than 798; result, a translucid enamel, like pure Chalcedony.

795. Crystallized Phosphate of Lime, Apatite.—Fusion attended with a phosphorescence and scintillation, but no decrepitation; result, a black shining slag, which, when filed, showed a globule of a high metallic lustre and unknown nature.

796. Phosphate of Lime of Estremadura, Compact Apatite.—Fusion easy; result, a white enamel.

797. Granular Sparry Phosphate of Lime, Apatite, from a matrix of magnetic iron oxide.—Result of fusion, a chocolate-brown glass, magnetic. The magnetism and colour owing to iron.

798. Pure Oxide of Magnesium, Magnesia.—Fusion, per se, extremely difficult; but moistened with water, desiccated, and placed on charcoal, the fusion was attended with a purple-coloured flame; result, a whitish glass, with a pseudo-metallic lustre.

799. Hydrate of Magnesia, Pure Foliated Magnesia from America.—Fusion incomparably difficult, attended with a purple coloured flame; result, a white opaque enamel, with a coat of limpid glass.

800. Iconite, Pagodite of China.—Fusion easy; result, a brilliant limpid glass.

801. Comolite, Potstone.—Fusion easy, attended with combustion; result, a dark green glass, containing minute acicular crystals.

802. Talc, all the foliated varieties.—Fused; result, a greenish glass.

803. Serpentine, most varieties of.—Fused; result, oak-apple-green coloured globules, with indented surfaces.

804. Pure Oxide of Aluminum, Alumina.—Fusion easy; result, a snow-white opaque glass.

805. Crystallized Oxide of Aluminum, Sapphire (pure blue).—Fusion easy; result, grotesque greenish glass balloons.

806. Crystallized Red Corundum, Oriental Ruby

—Fusion extremely rapid, attended with a variety of changes in colour and form; at the first application of the fusion, the mineral became a liquid like oil, which, when cold, became a white and opaque globule; this, after a third melting, gave a pink coloured bead. It was supported on charcoal.

807. Common Corundum, Greenish-grey Crystallized Primary Corundum, from the East Indies.—Fusion rather difficult, attended with a greenish flame, and an escape of gas; result, a greenish translucid glass.

808. Sub-Sulphate of Alumina, Alumina of Newhaven.—Fusion very rapid, attended with a partial combustion; result, a pearl white translucid enamel.

809. Wavellite.—Fusion easy; result a snow-white enamel.

810. Red Siberian Tourmaline, Apyrous Tourmaline, Rubellite.—Fusion attended with loss of colour; result, a white enamel; sometimes a limpid glass.

811. Andalusite, Apyrous or Infusible Feldspar of Haüy.—Fusion easy; result, a snow-white enamel.

812. Cymophane, Chrysolite, Chrysoberyl.—Fused; result, a pearl-white enamel.

813. Cyanite, Sappare, Disthéne.—Fusion ready; result, a snow-white frothy enamel.

814. This mineral was used by Saussure as a supporter to the common blowpipe, on account of its refractoriness.

815. Hyperstene.—Fused; result, a jet-black highly-lustrous glass bead.

816. Zircon, Jargoon.—Fusion very difficult, attended with a partial (internal) infusibility; result, a mass, the superficies of which exhibited a white opaque enamel.

817. Spinelle, the Spinelle Ruby.—Fusion easy, attended with combustion and volatilization; result, a loss of colour and weight.

818. Automalite, Spinellane, Zinciferous Corun-

dum.—Fusion attended with intumescence, and deposition of yellow oxide; result, a grey enamel, with a surface showing frost-like crystals.

819. Topaz.—Fused; result, a white enamel, covered with minute limpid bubbles.

820. Pycnite, Red Schorl, Schorlite, Schorlaceous Beryl.—Fused; result, a snow-white enamel.

821. Pure Precipitated Silica, Peroxide of Silicium?—Fusion instantaneous, attended with a copious discharge of gas; result, an orange-coloured transparent glass. It was supported on charcoal.

822. " As to the real nature of Silica, very little satisfactory information has been obtained: after a century spent in constant experiments for ascertaining the real history of this extraordinary combustible, Chemists remain ignorant whether it be really a metallic body, or a combustible resembling boron and carbon."—Dr Clarke.

823. Hydrate of Silica, Santilite, Pearl Sinter from Tuscany.—Fusion perfect, attended with dense white fumes; result, a translucid pearly enamel, globular, and bubbly.

824. Hydrate of Silica, Hyalite, a diaphanous assay.—Fused; result, a snow-white frothy enamel, bubbly.

825. Hydrate of Silica, Sand Tubes of Drigg, in Cumberland.—Fusion instantaneous; result, a pure limpid glass, bubbly.

826. Hydrate of Silica, Opal.—Fusion perfect; result, a pearl white enamel.

827. Hydrate of Silica, Chalcedony.—Fusion perfect; result, a snow-white enamel.

828. Hydrate of Silica, Egyptian Jasper.—Fusion easy; result, a greenish bubbly glass. It was supported in a platinum crucible.

829. Hydrate of Silica, Common Flint.—Fusion rapid and perfect; result, a snow white frothy enamel.

830. Crystallized Silica, Rock Crystal, a highly diaphanous specimen. — Fusion difficult, attended

with no loss of transparency; result, rounded vitreous masses, like Rupert's drops, bubbly.

831. Common White Quartz.—Fusion rather easy;
result, same as rock crystal (830).

832. Leucite, Amphigene, White Garnet of Ve-
suvius.—Fused; result, a perfectly limpid glass,
bubbly.

833. Peruvian Emerald.—Fusion easy, attended
with a deprivation of colour; result, a very limpid
hubbly glass bead.

834. Siberian Beryl, Aigue Marine, Asiatic Emerald.—Fused; result, a limpid bubbly glass.

835. Lazulite, Lapis Lazuli.—Fused; result, a
transparent greenish bubbly glass.

836. Gadolinite, Ytterbite.—Fusion rapid; result,
a jet-black highly-lustrous glass.

837. Pure Oxide of Plutonium or Barium, Barytes,
Barytic Earth.—Fusion very ready, accompanied by
a green flame, slight scintillation, dense white fumes,
and deposition of white oxide; result, a jet-black
shining slag, inclosing a regulus of Barium.

838. When the slag was cut by the file, a regulus was
disclosed, having the metallic lustre of silver. When cast
into water, gaseous bubbles were evolved, until the whole of
the metal, by decomposing the water, was again converted
into Barytes. Other properties, characteristic of the metallic
base of barytes were also exhibited. Sir Humphrey Davy,
to whom the metal of barytes, thus produced, was transmitted in naphtha, admitted the fact of its revival. The
metal is described fully in Dr Thomson's Chemistry.

839. Pure Oxide of Strontium, Strontian Earth.—
Fusion difficult, accompanied by an amethystine
flame, scintillation, dense white acrid fumes, and deposition of white oxide; result, a jet-black shining
slag, with internal metallic lustre.

840. Siliciferous Oxide of Cerium, Cerite.—Fusion
easy; result, a metallic bead, with a surface of shining acicular crystals.

841. Ferriferous and Manganesiferous Oxide of
Columbium, Tantalite.—Fusion instantaneous; re-

sult, similar in appearance to that of barytes (837).

842. Ferriferous Oxide of Chromium, Chromite.—Fusion easy ; result, a dark globule, with no metallic lustre, but magnetic.

843. Geniculated Oxide of Titanium, Titanite.—Fusion difficult ; result, a metal, exhibiting the lustre and colour of iron. It was supported in a charcoal crucible. With borax, ebullition took place, and the formation of acicular reddish crystals of oxide.

844. Oxide of Uranium, Pechblende, Uranite.—Fusion difficult, accompanied by loss of colour, smell of sulphur, ebullition and scintillation ; result, a brown globule, metallic lustre, brittle, hard, not magnetic. It was supported in a charcoal crucible.

845. Sulphuret of Molybdenum, Molybdena.—Fusion instantaneous, accompanied by dense white fumes, and a deposition of oxide, mixed with minute silvery metallic globules ; result, a metallic mass, resembling arsenical iron.

846. Peroxide of Scheelin, Tungstic Acid.—Fusion accompanied by combustion, volatilization, and a deposition of a deep blue and yellow oxide ; result, a bright copper-coloured metallic coating on the charcoal.

847. Ferriferous and Manganesiferous Oxide of Scheelin, Wolfram.—Fusion ready, accompanied by ebullition ; result, a bead like magnetic iron ore, lustrous, but not magnetic.

848. Metalloidal Oxide of Manganese, crystallized in prisms, Purest Ore of Manganese.—Fusion instantaneous, accompanied by combustion and vivid scintillation ; result, a metal, as white and brilliant as silver.

849. Grey Oxide of Manganese, (after ignition in a crucible).—Fusion quick ; result, a metallic slag, with dark coloured metallic lustre.

850. Carburet of Manganese, the substance which

floats on fused pig-iron.—Fusion accompanied by a brilliant scintillation, resembling the fire-work called a " flower-pot;" result, a bead with a bright metallic lustre, magnetic.

851. Black Oxide of Cobalt.—Fusion accompanied by a deposition of oxide, resembling Brunswick-black varnish; result, a metal, silvery, ductile.

852. Crystallized Sulphuret of Zinc, Resin Blende.—Fusion accompanied by a sapphire-blue flame, volatilization, and deposition of a white oxide; result, a metal, discernible in the centre of fused ore.

853. Nickel, alloyed with Palladium. — Equal bulks placed together on charcoal, easily formed a malleable alloy, surpassing in lustre the most splendid metals known; fit for telescope mirrors.

854. Nickel alloyed with Iron, equal parts by bulk.—Fusion accompanied by vivid combustion, previous to their union; result, a globule of white and highly-splendid alloy.

855. Tin Oxide, Wood Tin.—Fusion accompanied by a violet flame, scintillation, dense white fumes, and deposition of white oxide; result, a jet-black slag, not metallic, though with metallic lustre.

856. Granular Tin Oxide of the Molucca Isles, Tin-stone in grains.—Fusion easy, accompanied by a violet flame; result, tin in a perfectly malleable state. It was supported on charcoal.

857. Red Iron Oxide, Fibrous Red Hæmatite, Wood Iron.—Fusion rapid, accompanied by combustion and scintillation after reduction; result, a bead, nearly malleable, with metallic lustre. It was supported on charcoal.

858. Iron Wire and Steel.—Fusion accompanied by a brilliant and beautiful scintillation; result, a shower of minute globules of oxide of iron.

859. *Combustion of Iron Wire and of Steel.*—This affords one of the most brilliant and beautiful experiments with the Gas Blowpipe. Very stout iron-wire is consumed almost in the instant that it is brought into the gaseous flame; and

A A

its combustion is attended with such a vivid scintillation, that it displays a very pleasing fire-work. A part of the metal remaining fused at the end of the wire, is rendered brittle by the operation. If a steel watch-spring be substituted for the iron-wire, the effect is yet more striking; the combustion of the steel literally causing a shower of fire.

860. Atmospheric Iron Ore, Meteoric Stones.— Fusion momentary, accompanied by ebullition and combustion; result, a dense bead, magnetic, and with metallic lustre, but not quite reduced. It was supported on charcoal.

861. Copper Wire.—Fusion rapid, accompanied neither by ebullition nor combustion.

862. Alloy of Copper and Tin, Antient Bronze.— Fusion rapid, accompanied neither by ebullition nor combustion.

863. Copper alloyed with Zinc, Brass.—Fusion accompanied by a green flame, flashes of light, sputtering noise, and deposition of white oxide of zinc.

864. Ores of Silver, and Pure Silver.—Take fire, burn with a light green flame, the metal being instantaneously sublimed in dense white fumes.

865. Silver with the Metal of Barytes, formed an alloy which retained a metallic appearance for two months; but then, upon exposure to atmospheric air, assumed an earthy form.

866. Combustion of Pure Gold.—The metal was made to adhere, by borax, to a tube of pipe-clay. Upon exposure to the gaseous flame, the gold was volatilized, the pipe-clay partially fused, and the borax exhibited a beautiful glass of gold. The end of the tube was coated with the metal, and surrounded by a circle of oxide, of a most exquisite rose colour, exhibiting a striking appearance.

867. Arenaceous Ore of Platinum, Platina.— Fused; result, a globule of brittle alloy; aspect dull.

868. Pure Platinum.—Fusion very easy, accompanied by boiling, combustion, and scintillation; re-

sult, a mass of platinum. It melted as easy as lead does in a common fire.

869. Platinum alloyed with the metal of Barytes. —One grain of each was supported together, in a charcoal crucible; the result of the fusion was a bronze-coloured metallic alloy, which weighed two grains, and fell to powder in 24 hours.

870. Platinum with Silver.—The alloy was easily formed on charcoal; extremely malleable; lustre equal to pure silver; harder than silver; might be useful in the Arts or in coinage.

871. Platinum with Gold.—The alloy was formed easily on charcoal; colour and specific gravity according to the proportions of the two metals.

872. Platinum with Copper, equal parts by weight —Fusion easy, accompanied by a vehement ebullition; result, an alloy remarkably fusible; soft; malleable; in colour and specific gravity resembling gold.

873. Platinum with Iron, equal parts by weight. —Produced in a charcoal crucible, a combustion like a brilliant fire-work; the result was malleable, extremely hard, and highly lustrous.

874. Platinum with Iron, in equal parts by bulk. —Result, an alloy, brittle, and exhibiting a minute but brilliant crystallization.

875. Pure Palladium.—Fusion easy, accompanied by combustion and scintillation; result, a globule with a tarnished aspect.

876. Palladium with the metal of Barytes.—The latter was supported on the former, and the result was an alloyed surface, resembling bronze varnish.

877. Palladium with Copper, in equal parts by bulk.—Fusion rapid; accompanied, after union, by scintillation; result, an alloy, pale coloured, highly lustrous, very soft, very fusible.

878. Brittle Regulus of Rhodium, after purification by a process in the humid way.—Fused; result, pure rhodium, perfectly malleable.

879. Pure red-coloured Muriate of Rhodium.—
Fusion easy, accompanied by combustion, veheme
ebullition, and volatilization; result, a brilliant m
tallic globule, malleable. It was supported in
charcoal crucible.

880. Very Pure Granular Ore of Iridium.—Fusi
accompanied by ebullition, scintillation, and depos
tion of reddish oxide; result, a vitreous oxide of i
dium. It was supported in a plumbago crucible.

881. Granular Ore of Iridium and of Osmium.—
Fusion difficult, accompanied by combustion an
volatilization; result, a metallic globule exceeding
hard; very lustrous.

882. Crystallized Carbon, Amber-coloured Di
mond.—Became limpid; colourless; white; opaqu
diminished in bulk and weight; was covered wi
bubbles; shone with a metallic lustre; and was
volatilized in three minutes. It was supported in
charcoal crucible.

883. Carburet of Iron, very pure Plumbago.—
Fusion exceeding difficult, accompanied by viv
scintillation, but no change of colour; result, pr
duction of many very minute dark-coloured gla
globules, of an unknown nature.

MR GURNEY'S EXPERIMENTS WITH H
OXY-HYDROGEN BLOWPIPE.

884. I have never yet submitted any substance
the action of the instrument, which did not in son
way or other give way before it.

885. *Gun Flints* fuse instantly, and run into tran
parent glass—like stones. They should be calcin
prior to being submitted to the blowpipe, otherwi
their water of crystallization occasions them to
into small pieces.

886. *Porcelain* of all kinds fuses readily; exhibi
ing a beautiful opaque vitrification.

887. *Common China* melts into perfect crystal.

888. *Tobacco Pipes* fuse easily, and become vitrified, assuming a yellow colour.

899. *Rock Crystal* fuses and gives out considerable light during the operation.

890. The *Diamond* burns and becomes dissipated instantly, producing a fine orange-coloured light.

891. All the *Precious Stones* melt, generally, into transparent substances.

892. The *Earths* are all strongly affected by its intensity, and exhibit very striking and beautiful phenomena :—Magnesia throws off brilliant sparks, and fuses into hard granular particles, which scratch glass.

893. All the *Metals* fuse, and, if the operation is prolonged, inflame.

894. A large *Steel File* burnt in a most astonishing manner; in fact, it produced a brilliant firework; the globules first separated were thrown several feet in the atmosphere, where they created the most beautiful scintillations.

895. Steel and Iron in every form, also, Gold, Silver, and Copper, melt and burn very easily; assuming vivid colours. Platinum fuses and scintillates.

896. Stones, Slates, and Minerals of every description, melt, or are volatilized. Indeed, they appear to be torn into pieces, as it were; their elements subliming and forming new combinations with each other.

897. By mixing with Gun Flints, &c. in a state of fusion, metals and other substances capable of imparting vivid colours, very beautiful artificial gems may be formed.

898. The light produced by burning Magnesia is so powerful, that the eye can scarcely look on it; while that accompanying the inflammation of pure Lime is so intensely brilliant, as to be absolutely insupportable. It resembles day-light, and throws

all other artificial light into shade. One of our
largest theatres might be illumined by it with splendid effect ; and, however fanciful the idea may be,
I cannot help thinking, that light produced in this
manner, might be used with great advantage in light-
houses.

899. I have melted by this instrument a platinum
bar a quarter of an inch in diameter ; and have kept
six ounces of that metal in fusion at one time by
the jet.

CHEAP TABLE BLOWPIPE.

900. The construction of a temporary blowpipe to
supply, in cases of necessity, the place of a table
instrument (45), is not difficult. A piece of glass,
pewter, or any other pipe will convey the air, and
being tied to a weight or stand, or even a candle-
stick, may be arranged at the proper height, that its
jet may accord with the lamp to be used. The first
rough jet may be made by drawing out a piece of
small glass tube in the spirit lamp, or a candle, and
being attached to the apparatus, a second and proper
jet may be made by means of it, out of a thicker
piece of tube, and substituted for the smaller one.
Instead of the bellows a large bladder may be used,
or what is better, a bag made of oiled silk, or some
of those fabrics now sufficiently common, in which
cloth is rendered air tight by caoutchouc. This may
be placed under one end of a board with weights
upon it, or within a portfolio, subject to pressure,
and the air may be thrown into it from the lungs
by another piece of tube sufficiently long to reach
to the mouth. This tube will require a valve to
prevent the return of the air, and the simplest that
can be constructed for the purpose, in an extempo-
raneous manner, is perhaps the following :—

901. A piece of any tube of about the diameter of

that represented by fig. 51., having a smooth and
level end, is to be selected, and also a strip of black
oiled silk of a width rather more than the external
diameter of the tube, or if that be not at hand, a
piece of ribbon of the same width, which has been
rubbed with wax, so as to have the interstices in it
filled up without destroying its flexibility.   This is
to be adjusted loosely over the end of the tube, and
the extremities folded down on opposite sides, and
tied with a piece of thread; the silk itself not being
so tight, but that by applying the mouth to the
opposite end air may be easily blown through the
tube, and out at the extremity between it and the
silk; and yet so near that a pressure being exerted
in the opposite direction, the silk will be carried
against the end of the tube, and prevent the air
from passing that way.—The tube by which air is
to be thrown into the bag from the mouth, is to
have such a valve constructed at its extremity, which
is to be introduced through a hole made in the bag,
and tightly tied in it by a few turns of twine. So
arranged, the bag is easily filled by air from the
lungs, which being gradually expelled at the jet,
gives energy to the flame. All the tubes required
for this instrument, except the jet, may be made
even of paper.

902. The elasticity of writing paper is such, that a tube
constructed of three or four evolutions, being tied round
with twine, and preserved from rough treatment, may be
considered as air tight at low pressures. When the edges
of the paper are pasted, the tightness is more permanently
insured. If the whole of one side of the paper be pasted
before it be rolled up, the tube will be still tighter and
stronger, and being then varnished or covered with a coat
2, drying oil, will serve for the conveyance of steam and
water, as well as of gases. Waxed paper also makes excellent tubes for similar purposes. The tubes which are
pasted or cemented together, are easily made by folding
them upon a round ruler or wooden rod, or upon a thick
wire, a piece of loose paper having been first put upon the
rod, to permit of the tube's ready removal when finished.

903. A lamp for such a blowpipe is soon fitted up; a bundle of cotton threads placed at the side of any small vessel filled with oil, will answer the purpose, and none is more convenient than a little Wedgewood or evaporating basin.—*Faraday on Chemical Manipulation.*

HOW TO HOLD THE BLOWPIPE.

904. " The lamp or candle should be low, that whilst using the blowpipe the arms may rest steadily on the table. The hand should lay hold of the instrument as far from the mouth as convenient, greater freedom of motion being thus obtained, and the left or the right hand should be used indifferently. Steadiness is requisite, and it should be the constant endeavour of the student when he is using the instrument, to obtain the power of so apportioning his breath, and of retaining the instrument and the flame, that the latter shall appear like a fixture, and neither change in appearance nor direction for several minutes together; yet this with such lightness of touch and easy hold, that he may at pleasure send the flame in any direction, and upon any place he pleases."—*Faraday.*

TOFT'S BLOWPIPE.

905. This instrument was first described by Dr E. D. Clarke, in an article, in the Annals of Philosophy, "Upon a New Hydro-Pneumatic Blowpipe, so constructed as to maintain during two hours, uninterruptedly, a Degree of Heat capable of melting Platinum; and this by propelling the Flame of a small Wax Taper with Atmospheric Air."—

906. "The advantages of the old instrument (the Common Hydro-Pneumatic Blowpipe), consisted in the operator having both his hands at liberty; and in the relief which it afforded from that fatigue and possibility of injury to the lungs incident to a protracted restraint on their free action, to which persons using the common mouth blowpipe were liable. To these advantages, which the new instrument also possesses, we may add the following:—

907. "Either common air, or any other gaseous fluid may be used for the propelling current, by condensing it in the reservoir, and thus experiments may be made on the fusing powers of the different gases with perfect ease and convenience.

908. "The power of entire exhaustion possessed by the new instrument, ensures the operator from any admixture of common air, where *oxygen gas*, or any other gaseous fluid is to be employed.

909. "The old instrument, although very useful for bending tubes, or other ordinary purposes, required to be repeatedly restored to action by fresh supplies of air, at intervals seldom exceeding five minutes, in the common-sized instruments. In the new instrument there is this great improvement; that a steady flame of two hours' continuance may be maintained, of the most perfect shape and uniform temperature, uninterrupted by casual currents from the pneumatic reservoir.

910. "The troublesome interruptions caused by the ejection of water, while supplying the apparatus with air, which were common in the old instrument, do not happen in the new one.

911. "The new instrument may remain unemployed for any length of time, being always ready for instantaneous use, and requiring no other preparation than merely that of lighting the wax taper employed to supply the flame.

912. "The manner in which this instrument wa
brought to its present state of perfection, affords a
anecdote which may not be uninteresting to th
reader; because it will show the force of mechani
cal skill, as it is sometimes remarkably conspicuou
in uneducated minds. A servant of mine, who ha
been frequently employed in attendance during m
lectures in mineralogy, seeing me reject the old in
strument as unfit for the uses to which I wished t
apply it, asked me the reason of my setting it aside
To this I answered that the short space of time i
which it continued to propel the flame rendered i
inconvenient; and that I would rather use the com
mon mouth blowpipe than be liable to such frequen
interruption from the necessity of supplying it ever
five minutes with fresh air. This caused him t
inspect the inside of the instrument; when he simpl
observed, *"It is very awkwardly contrived; I coul
make a better myself!"* and in good earnest, with
out further communication with me upon the sub
ject, he fell to work, and produced the new improv
ed apparatus which it is my present purpose to de
scribe, and to make as generally known as possible
The inventor's name is *Johnson Tofts.* Upon th
principle which he has adopted for the improvemen
of his blowpipe, such instruments are now manufac
tured in London. By means of one of these instru
ments, gold might be exhibited in a state of con
tinual fusion for almost any length of time.

913. "A trial has been made of the use and powe
of this improved blowpipe throughout an entire cour
of public lectures in mineralogy before the Univer
sity, with such success, as to produce sufficient proo
of its convenience and efficacy. The effects witness
ed when oxygen gas was employed exceeded tho
usually produced by the same agent from a gasom
ter; owing to the condensed state in which the g
was propelled; while the cone was much more ma
ageable. Platinum wire of some thickness was fused

platinum foil offered no resistance whatever. The steel mainspring of a watch underwent brilliant combustion. And even when the instrument is charged with common atmospheric air, thin cuttings of platinum foil sustain an instantaneous fusion.

914. " Had we possessed this apparatus before the gas blowpipe was invented, many of the results obtained by that powerful instrument would have been anticipated. It is not to be expected, however, that the fusing powers of the two blowpipes can be compared together; but from the safety of Toft's blowpipe, a child may use it; whereas the other would be indeed a dangerous toy."

915. The diagram in the margin will exhibit the construction of this improved hydro-pneumatic apparatus. When the instrument is charged for use, A A is filled with air, and the water remains above it in the vessel B, whence it will descend through the cylinder O so long as any air continues to escape from the reservoir A A through the pipe J, to which the blowpipe jet is affixed by a stop-cock. When the lower reservoir contains no air, it will be entirely filled by water; so that whatever air or gas it may be desirable next to introduce will again displace the water, and drive it up to B, and occupy its place, without receiving any admixture of common air.—All gases are introduced by means of a bladder and condensing syringe, which screws by means of a stop-cock to the pipe I. And it has been found better to introduce atmospheric air with a syringe than to fill the reservoir with air from the lungs: yet, air may be blown in from the mouth if a syringe be not at hand.—The machine is supplied with water, or the

water is removed from it by means of a common syphe
—The usual size of one of these blowpipes is two fe
high, two feet long, and five inches wide. They a
made of copper or tin, and enclosed in a wood
case, which serves as a table, and a rest for t
arms ; the wax taper, being sunk into a cylinder
C, is elevated or depressed by means of a screw a
a rack.   But a stationary spirit-lamp, if it should
preferred, may be fixed in the same place.   T
cylinder O should be three inches in diameter, a
reach to within half an inch of the bottom of t
vessel.   The dotted line at 12 inches shows t
height of the water. At D is a small aperture f
the admission of atmospheric pressure.   The i
strument is so simple that a more minute descri
tion of it is unnecessary.

PHILLIPS'S SPIRIT LAMP.

916. Every student should be acquainted w
this modification of that useful instrument,
spirit lamp ; the advantages of which consists
its ready construction in any place and by any pe
son.—Let a piece of tin plate about an inch lo
be coiled up into a cylinder of about three-eighths
an inch in diameter. If the edges are well ha
mered, soldering is not necessary.   Perforate a co
previously fitted to a phial, put the short tin tu
through the cork, and a cotton wick through t
tube: the lamp is now complete, and capable
affording a strong flame. Care must be taken
to prevent the rise of the spirit by fitting the co
too closely. A cover for this lamp, to prevent t
evaporation of the alcohol when the instrument
not in use, may easily be made of a piece of gl
tube closed at one end.

INDEXES.

*** *The References are to the* PARAGRAPHS *of the Work,
not to the* PAGES.

I. GENERAL INDEX.

B B

II. INDEX TO THE MINERALS.

c c 2

III. INDEX TO THE PLATES.

THE END.

MALCOLM & GRIFFIN, PRINTERS, GLAS